D0915427
DISCARD

Troubled Skies, Troubled Waters

Troubled Skies, Troubled Waters

THE STORY OF ACID RAIN

Jon R. Luoma

The Viking Press
New York

First published in 1984 by The Viking Press
40 West 23rd Street, New York, N.Y. 10010

Published simultaneously in Canada by
Penguin Books Canada Limited

Portions of this book appeared originally in *Audubon, MPLS St. Paul* magazine, and *The Sciences* magazine.

LIBRARY OF CONGRESS CATALOGING IN PUBLICATION DATA

Luoma, Jon R.
Troubled skies, troubled waters.
Includes index.
1. Acid rain. I. Title.
TD196.A25L86 1984 363.7'394 83-47927
ISBN 0-670-73263-X

Printed in the United States of America
Set in Times Roman

For my father

Despite major research efforts by a large group of concerned individuals and agencies, the evidence of widespread damage from acid rain remains less than convincing.

ALAN W. KATZENSTEIN,
public relations consultant
to the electric-utility lobby

It's hard to argue with a dead fish.

GEORGE HENDREY,
acid-rain researcher

PREFACE

This is a story about the unthinkable.

A vast tract of wilderness land lies along the northern border of Minnesota and the adjoining border of Ontario. Formally designated an undeveloped, primitive region by both the United States and Canada, the place is virtually roadless, yet there are highways of a sort here—ancient routes through a galaxy of lakes linked by thousands of miles of streams, trails, and portages. The primary mode of travel here is the canoe, that old conveyance of the Ojibwa and the fur-trading voyageur.

On the Canadian side of the border, the region is called the Quetico Provincial Park; its one million acres represent the third largest of the huge Ontario wilderness parks. In the United States, the Boundary Waters Canoe Area (BWCA) is the only surviving lakeland wilderness and the largest wilderness region in the eastern half of the country. But despite whatever dotted lines are imposed on the maps of nations, the Quetico and the BWCA are a single, intact ecosystem—more than two million acres of lake, river, pine forest; a land of eagle, moose, and timber wolf; a piece of nature sometimes called, simply: canoe country.

Here is the grimmest of ironies. Canoe country is defined by its water and its rock. The water here is often tannin-stained, the color of weak tea, but it is so clean that you can look straight down, fifteen feet or more, to a lake floor littered with chunks of glacial rubble. On some of these lakes, you can float through channels among entire archipelagoes of rocky islands—gently beveled mounds of lichen-carpeted bedrock where pine and birch cling to an impossibly thin pancake of soil. These granite islands and outcrops are, in fact, some of the oldest exposed rock to be found on earth.

Canoe country lies hundreds of miles from the industrial heart of the continent, yet it appears that this seemingly isolated region's water and rock make it highly susceptible to a bizarre form of ecological destruction. It appears, in fact, that at least part of the canoe country may be dying from industrial pollution.

Already, in other parts of the northern hemisphere, lakeland ecosystems that look strikingly similar to canoe country have been poisoned: most of their life is literally sterilized out of them. In North America alone, hundreds of lakes have already been turned into what a Canadian documentary film calls "wet deserts."

This story begins and ends in canoe country because its lakes have so far survived, and because there is reason to hope that these wilderness waters and tens of thousands of others like them can still be saved.

ACKNOWLEDGMENTS

Beyond acknowledging everyone quoted or mentioned herein for assisting in compiling information for this book, I would especially like to thank Gerry Foley of the Canadian Consulate in Minneapolis, Eville Gorham of the University of Minnesota, and Harriet Stubbs of the Acid Rain Foundation. Special thanks to J. David Thornton for review of the manuscript for errors of fact and to Gary Soucie, Roxannna Sayre, Les Line, and the rest of the editorial staff of the outstanding nature and conservation magazine *Audubon,* for funding the research for, editing, and publishing an article that would later grow into this book. Profound thanks to Don Demarest, Marian Young, William Strachan, and Pam Hendrick for wise counsel, persistence, faith, and tolerance.

Troubled Skies, Troubled Waters

I

I wake at some hour after midnight, jarred out of a deep sleep by the racket outside our tent. Groggy, I imagine for a few seconds that it is a dog barking nearby, then realize there are no dogs for miles and that the noise is the raucous, repetitive hoot of a barred owl: a series of cellolike hoots, a pause, then just a trace of an echo. Another hoot-hoot-hoot-hoot, another pause, another faint echo.

There is a long silence. No owl, just the soft rustle of wind in the pines. I wait. The silence prevails. Then, suddenly, from far across the stretch of Rose Lake, a loon begins its long, hollow, mournful territorial yodel. The bird's eerie cry is tentative at first, but soon it is blasting like a foghorn in the night.

A common loon is not a large bird—bigger than a duck and smaller than a goose—but it virtually bellows its various calls, and no other northwoods beast has a cry so otherworldly, yet so strangely human. More than one new visitor to the northwoods, awakened by the loon's wild laugh, has been instantly, blood-chillingly convinced there is a madman somewhere in the pines nearby.

The loon falls silent; the owl resumes. Silence. Then, in sudden counterpoint, the nearby owl and faraway loon are barking and bellowing their songs in weird and sublime unison.

It is spring in the canoe country. Ice is off the lakes only a month now, and outside my cocoon of sleeping bag, the night air is still sharp-edged on my face. The bird racket has woken my wife, Pamela, too, and she turns and mumbles something about owls. I fold back the tent flap. I can't pick out an owl's silhouette in the branches of pine and birch that rim the slab of bedrock we're camped upon, but I do see our Duluth pack hanging between two trees near the water: our food cache, out of the reach of foraging bears.

Beyond the suspended pack lies the glassine surface of Rose Lake. The loon is out there, hollering the maniacal cry that is the very signature of this land, for the gentle bird breeds only in the north and flourishes only in the wild. On this spring night, our loon calls out in virtually the same way its own distant ancestor called to the Indian hunting parties and French fur traders who passed here before.

Rose Lake, long and narrow, is one in a string that in the nineteenth century served as a link between the settled, eastern half of North America and the great boreal forest and tundra of the continent's far north and west. The Indians traveled here by birchbark canoe, as did the French fur traders, who in the mid–eighteenth century penetrated the border lakes just inland from the north shore of Lake Superior and before long learned that these linked, labyrinthine waters could serve as a complex of highways. The voyageurs came from Montreal, up the Ottawa and Mattawa rivers, through Trout Lake and Lake Nipissing, the French River and Georgian Bay on the northeast shoulder of Lake Huron; then past the long rapids of St. Mary's River where now stand the locks of Sault Sainte

Marie; and finally through the 450-mile passage along the north shore of the greatest of the lakes, Superior, until at last they reached the settlement at Grand Portage, just south of the Pigeon River, which now serves as part of the U.S.-Canada border.

At Grand Portage the voyageurs switched from the long Montreal canoes that had carried them across the Great Lakes to the smaller north canoes—vessels twenty-five feet long that could carry six to ten men and hundreds of pounds of provisions and trade goods—for the grueling, ascending nine-mile carry around the Pigeon River rapids. At the end of that long, rocky portage, the voyageurs reached what is now the canoe country, then traveled 200 miles through a chain of lakes and across thirty-six more portages to the big waters of Rainy Lake, and another 230 miles to Lake Winnipeg, whence the voyageurs could trap and trade west as far as Oregon and north into the subarctic tundra.

Rose Lake lies directly on that old route from Lake Superior to the interior. In late summer, canoes laden with trade goods would pass this very spot bound for the trapping and trading grounds. By spring, the traders would be heading back with their tons of fur, to converge on Grand Portage for a massive annual conclave of *hommes du nord* in July. There are those who say their ghosts are out there still, their long canoes gliding along the lakes.

Perhaps they stopped here on their journey to the annual conclave at Grand Portage, at this spot where Pam and I and our companions Lee Eddison and Jeff Madsen are camped. Surely, of the thousands of Indian and voyageur parties who passed through this slim bottleneck of a lake, some bivouacked on this same spit of glacier-scarred rock. Quite probably, a few woke groggy to the hooting and sawing and tremolo laugh of an owl and a loon in concert.

I look out of the tent again and realize, for the first time, that there are clouds. For two days, the sky had been clear; the stars had been sharp, white asterisks in the cold night air as we sat beside our campfire. Carelessly, we have made little provision for rain.

When I wake again, it is just dawn and raining lightly. Through the green translucence of our nylon tent, I can see water drops, first clinging to, then sliding down the fabric. Within minutes, there is a strong, steady downpour—not a raging thunderstorm but a hard, driving rain—and outside the tent, water is pooled already in the hollows of rocks and streaming from the leaves of trees. The only reasonable course of action is to go back to sleep.

We break camp after a pitiful breakfast of instant coffee and dried fruit, rolling our soaked-through tents into canvas packs and unslinging our food pack, which is heavy already with the weight of accumulated water. We load the canoes and push off. The rain continues, pools of water collect in the hulls, and our khaki packs turn dark brown.

We paddle to the west, covering the old voyageurs' water. Should we continue westward for a few more hours, we'd cross the Rat Portage, between Rose Lake and South Lake, and just after that the Height-of-Land Portage to North Lake. Height-of-Land held special significance for novice voyageurs. A mere quarter mile long, it climbed the hydrologic divide between waters draining east into the Great Lakes system and those draining north into Hudson Bay. Appropriately, it was at Height-of-Land that a novice was sprinkled with water from a cedar bough, presented with a plume for his hat, and, to salvos of gunfire, initiated into the fraternity of full-fledged voyageurs.

We instead turn to shore beside a high, slim waterfall that empties the waters of Duncan Lake, to the south, into Rose.

We are not elegant about beaching here; we simply steer the canoes reasonably close to shore and stomp into the water to unload. Jeff carries one canoe on his shoulders and I the other; Pam and Lee each shoulder a heavy pack filled with a wet tent and saturated provisions.

At Duncan Lake we lower the canoes to the water. I toss my rain poncho into the hull of the canoe, for the arms of my nylon jacket are already drenched with rain and the shirts beneath soaked with perspiration. By the time I make the return trip for the remaining packs, I'm wet fully through—jacket, shirts, trousers, socks, boots. The Duluth pack I lug back is wet. Wet white mist advances around the bluffs of Rose Lake behind me as I move up the portage. The rain forms rows of droplets like soldiers on the needles of the pines. Water gathers in the hollows of shrub leaves below until it turns the leaf downward to release a tiny torrent onto the thin soil. Water seeps through the soil to the bedrock two or three inches beneath, then obeys the pull of gravity and leaches slowly through the soil along the contour of impermeable bedrock, down the face of the bluff toward Rose Lake.

Everything here is water. Soaked to the skin, we paddle into the mist on Duncan Lake toward one short, easy portage into Bearskin Lake and home, and I suddenly feel a surge of joy at the unbounded wet beauty of it all, the misty cliffs and the steaming forest and the billions of dimples of rain-spatter on the lake.

It isn't until I'm driving toward Minneapolis that the intrinsic irony of it all hits me once again. The Indians lived in the canoe country, taking from nature nothing that was not somehow returned. The French and English voyageurs exploited the countryside and depleted the fur-bearing mammals through overharvest, but with a few notable exceptions, the mammals have survived. Later, poor logging practices savaged portions of what is now the canoe country, and there have been years of fire, disease, and drought, but the land has survived. In this century, there have been battles over damming the canoe-country waters and over banning commercial development and road building in the area. More recently, a heated dispute over restrictions on motorboats and snowmobiles in the area deeply divided the people of Minnesota, but the canoe country itself has endured.

So what confounds the imagination is that within five, perhaps ten, perhaps forty or one hundred years, we may no longer be able to listen to the call of the loon in the night beside some of these wilderness lakes; the trout may no more cruise beneath our canoes. Perhaps before too long, we'll return to a cherished lake to find that many of the creatures that once inhabited it and the surrounding forest are simply gone. Perhaps we'll see the same sort of destruction in the BWCA and the Quetico Park that others have witnessed in the Adirondacks, New England, Canada, Sweden, and Norway.

For it seems that the rain over the canoe country, the gentle essence of life, has become acid rain. The acid rains (and acid snows, mists, and sleets) are formed from air pollutants emitted miles away, and if that pollution is not soon controlled, there is little reason to question if the waters in the canoe country will become acidified, for the thin soils and unique bedrock geology that characterize this remarkable wilderness are much the same as those in other regions where lakes and streams have already been destroyed by acid, including Scandinavia, where as many as 20,000 lakes have been acidified to the point that no fish and few aquatic insects can survive, the Adirondack Mountains of New York, where more than 200 lakes are lost, and Ontario, where more than 100 lakes are functionally dead.

Unless action is taken in places as far away from canoe

country as Ottawa and Washington, D.C., Sudbury, Ontario, and Columbus, Ohio, we will be left only with the question of *when,* of how many years are left before these canoe-country lakes are lost, and the question of how many *more* lakes in those other places—Scandinavia, upstate New York, parts of Canada—will be destroyed.

Nearly three hundred miles south of canoe country, near the intersection of Interstate 35W and State Highway 36 in Roseville, a Minneapolis–St. Paul suburb, a brick office building sits just out of sight behind a motel. It is an unlikely place for an international controversy to begin.

The modern but thoroughly undistinguished structure is the headquarters of the Minnesota Pollution Control Agency (MPCA). There some three hundred engineers and technicians and assorted support staff "control pollution" in much the same manner as do their counterparts in other state and federal agencies.

The employees do not attempt to stop pollution, because that is impossible. Instead, they endeavor to set limits on the amounts and types of waste any industry or other entity may discharge into the air, water, or soil; specifically, they write permits that determine what a source may and may not discharge. Often, environmentalists complain that such permits are little more than "licenses to pollute," while industry lobbyists routinely insist that the permit limits are much too strict and that the staff is scheming to undermine the state's economy. But the engineers and technicians persevere. Because paper is the unavoidable offal of any regulatory beast, these people produce great piles of it in a valiant attempt to manage the unpleasant leftovers of an affluent, industrial society—its smoke, garbage, and sewage.

On the building's fourth floor, in the back of a file drawer in the Public Information Office, an old news release dated August 8, 1977, announces a visit of a top official of the U.S. State Department to the MPCA. The purpose of that visit was to discuss a developing, but still minor, conflict between the state of Minnesota and a foreign nation.

In 1977, Canada was planning to build a large electrical generating plant in the western Ontario village of Atikokan, just a few miles from Minnesota's northern border. In the months that preceded the State Department's visit, officials at the MPCA had become increasingly alarmed at this prospect.

On the face of it, Atikokan seemed an odd choice for the site of a huge power plant. Ontario is a vast province, encompassing 412,000 square miles (roughly the area of New York, Pennsylvania, Ohio, Michigan, Wisconsin, Illinois, Indiana, and all six New England states), but the great bulk of Ontario's population of eight million lives in a narrow band between Lake Huron and the St. Lawrence River, with Toronto at the center. In fact, 85 percent of Ontarians live in an area that constitutes only 15 percent of the province's land.

Atikokan, a largely depressed mining village on a lonely highway, is in the western part of the province, one of the most sparsely settled chunks of territory in all of North America. The village is a thousand miles from Toronto, yet that was where the province-controlled electric utility, Ontario Hydro, proposed to build a huge, 800-million-watt coal-burning power plant, even though the utility already was producing 40 percent more electricity than its customers could consume.

Ontario Hydro appears to have chosen Atikokan as a site for the power plant primarily for political reasons. By the mid-1970s, the village's major employers, the Steep Rock Iron Mine and the mines and ore-sintering plant of Canadian Inland Steel Ltd., had pretty well exhausted Atikokan's iron-ore resources, and both operations were closing down. Economic prospects for the community were gloomy, but if the government were to build a huge new power plant, some one thousand contruction jobs would be created temporarily and a few hundred employees would eventually find permanent work running the plant. It was always questionable how many of those workers would actually come from the ranks of Atikokan's unemployed, but for a time, at least, money would pump into the village's economy. Shopkeepers and bankers could stay afloat; the schools would remain open.

A tiny handful of area residents objected to the prospect of the power plant, especially to the pollution that would come from the facility. Ontario Hydro did intend to install one type of air-pollution-control device—equipment to remove "particulates," or particles of dust and soot and ash—and the company planned to construct a tall smokestack to help carry any harmful emissions away from the village itself. But the utility refused outright to install equipment to remove the colorless pollutant gas, sulfur dioxide.

Atikokan lies just north of Quetico Provincial Park, yet Ontario Hydro insisted that the power plant's emissions of sulfur dioxide would have no effect on the sensitive plant and animal life of the wilderness region. The local critics wanted the government to assemble a team of experts to assess the possible environmental impact of the emissions, since such a full, independent assessment had never been made, but they were rebuffed. In mid-1977, the Ontario government gave the green light to Ontario Hydro to build the plant in Atikokan, agreeing with the utility's own assessment that any pollutant emissions would be insignificant.

Criticism of the project was unwelcome in the village's tightly knit commercial and political community, and to this day, one of those early dissenters insists that his name not be used in connection with his activities, for his own job has been threatened because of his involvement.

In 1976, frustrated by the provincial government's refusal

to respond to pleas for an environmental evaluation of the proposed plant, he drove from Atikokan to Roseville, Minnesota, to the headquarters of the MPCA, to beg the agency to conduct at least an initial analysis of emissions from the big plant. He pointed out that even though the coal burner was in Ontario, it would lie only thirty miles from the border, and on the United States side of that border lay the Boundary Waters Canoe Area wilderness and, just west of the BWCA, the new, canoe-oriented Voyageurs National Park, both prized for their beauty and their unique ecosystems.

Eventually, his complaints came to the attention of two air-quality analysts at the agency, Brad Beckham and Aaron Katz, who began to collect information about the sulfur content of the coal that would pour into the plant's boilers, the combustion efficiency of those burners, the height of the smokestack, the typical behavior of a stream of pollutant gas leaving such a chimney, and wind patterns in the region. With the aid of a computer, they began to prepare a model to discover just how much sulfur-dioxide gas could possibly blow from Atikokan into northern Minnesota.

In large doses, sulfur dioxide is known to cause serious respiratory problems in humans. There was little likelihood that such high concentrations could reach Minnesota from the tip of a 650-foot chimney thirty miles away, but Beckham, Katz, and others at the agency wondered if on at least a few days each year the Atikokan emission might violate special, strict U.S. standards set to protect the air in pristine areas such as the BWCA.

As they worked, Beckham and Katz began to develop a new set of concerns. They had heard rumblings in the scientific press of damage to lakes and aquatic life in some remote regions of New York State and in portions of Scandinavia, areas that are similar geologically to northern Minnesota. Scientists who had studied those damaged lakes were saying that the lakes were apparently being poisoned by precipitation—rain and snow that had become polluted by sulfuric and nitric acids. And those acids were forming as a result of sulfur-dioxide and nitrogen-oxide emissions from such sources as the proposed coal burner at Atikokan.

By the time the August 1977 news release was written, Beckham, Katz, and their superiors had become so concerned about this poorly understood form of pollution that they asked the U.S. State Department's Office of Canadian Affairs to intervene on behalf of the state of Minnesota.

"The MPCA," said the news release, "has expressed particular concern that acidified sulfur gases will enter the soil through rain and snowfall."

"We only had a gut feeling about it then," recalls Beckham. "It was all so new to us. There just wasn't enough information about acid rain."

At the time, the term was seldom heard outside a closed circle of atmospheric chemists and freshwater biologists. These few scientists were suggesting that acid rain first formed from the reactions of sulfur-dioxide and nitrogen-oxide pollutants in the atmosphere, then fell to earth, and then, in certain types of areas—that is, in regions characterized by thin, low-alkaline soils—destroyed lakes and rivers by disrupting their chemical balance.

Though acid rain (most scientists would prefer a more technically accurate term such as "acid deposition") was a poorly understood phenomenon in 1977, there was alarming evidence of its effect.

Around 1970, a New York State forest ranger named Bill Marleau, who held no degrees in ecology or atmospheric chemistry, began to notice subtle signs of disruption of the web of life near a remote mountain cabin he built and owns in the western Adirondack Mountains of New York. Marleau is a keen observer and a lover of nature, and for years a deafening choir of croaking and singing frogs had kept him awake in the summertime at his cabin. One year, the frogs stopped singing. Once, mayflies hatched in great fluttering blizzards over the surface of the small lake beside the cabin, but suddenly the mayflies vanished. Then there were no more trout in the lake, then no loons.

It made no sense. The lake appeared to be as clean as ever. In fact—and this was bizarre—the lake seemed even *cleaner* than ever. There was virtually no fishing done at the lake, no town or factory for miles upon miles. Yet this remote mountain lake was dying.

At about the same time, a University of Toronto biologist named Harold Harvey was conducting a simple experiment. He wanted to see if salmon could thrive in a small, but deep and cold, Ontario lake as they had thrived after successful attempts by the state of Michigan to introduce salmon in the Great Lakes. Harvey found a remote, sixty-foot-deep lake that was oligotrophic; that is, a "young" body of clean, clear water, not heavily enriched with fertilizing nutrients that promote the growth of algae. It seemed a perfect candidate for salmon. Previous government surveys had already proved that the lake supported a healthy population of native lake trout (which are also of the salmonid family) as well as herring, perch, and other species. The experiment seemed likely to succeed. Harvey planted four thousand salmon fingerlings, but when he returned a year later, he could net none. In fact, there were no trout, no herring, no perch, only a few white suckers that were oddly stunted, with flattened heads and twisted spines. When Harvey tested the water chemically, he found that it was highly acidic.

In August 1977, as the Minnesota Pollution Control Agency prepared for its meeting with the State Department official, acid rain hardly ranked among the leading environmental issues of the time: smog, hazardous chemical dumps, pesticides, nuclear radiation, garbage, and the most ubiquitous and most unglamorous pollutant of all, sewage. Yet, within three years, the leading environmental advisers to a U.S. president would be calling acid rain one of the greatest threats of a new decade. Ironically, in 1983, the Atikokan power plant would still be under construction (although reduced in size) but would be almost forgotten in a flurry of much more disturbing findings about this strange pollutant. In the end, roles would be reversed. *Canadian* officials would be demanding strict pollution control from coal-burning power plants in the United States to abate an across-the-border acid threat, and across the planet's northern hemisphere scientists would begin to find hundreds of remote, crystal-clear, and, by all appearances, unpolluted lakes and streams that are, in fact, dead.

2

Early one hot, sticky July morning in Columbus, Ohio, the telephone rang in my hotel room. It was Bill Vaughan, vice-president of the consulting firm Environmental Measurements, Inc.

"You might want to get over to Mission Control," Vaughan said. The PEPE had arrived. Vaughan's phone call would lead, in short order, to a firsthand look at acid rain in the process of being made.

A half hour later, I found Vaughan in the basement lounge of a dormitory at Ohio State University that was serving, for these few summer weeks, as Mission Control for a unique and elaborate air-pollution study that Vaughan and his firm were coordinating for the U.S. Environmental Protection Agency. The EPA, with the sort of hieroglyphic touch that could come only from a federal agency, had dubbed the program the "Summer 1980 PEPE/Neros Field Measurement Program." Its central purpose was to examine a "persistent elevated-pollutant episode": a PEPE. Dozens of separate entities were involved in the project, from the National Oceanic and Atmospheric Administration to Argonne National Laboratory to an array of private consulting firms that specialize in pollution studies. But more astonishing was the equipment that had been assembled for the study: a dozen aircraft and five mobile ground laboratories stuffed with sophisticated monitoring gadgets. In the dormitory lounge, teams of scientists clustered about chart-strewn tables, and a pair of two-way radios jabbered intermittently. In an adjacent room sat a scientific computer, moved here for this mission, and in another was a weather station linked by teletype to the National Weather Service.

Vaughan was standing before a huge map of central Ohio mounted on a billboard to one side of the lounge. "Look at this," he said, gesturing at the map. "It's just beautiful."

He was pointing at a series of plastic pushpins that formed a neat spiral, like an uncoiling spring, around the city of Conesville, Ohio. Vaughan's excitement was understandable, since this was precisely what he and the team headquartered in this dormitory had been waiting for. The pushpins were the footprints of a stagnant high-pressure weather system—the type of system that, among other things, can be particularly efficient at producing and transporting acid rain.

The project was an omnibus of air-pollution studies. Research airplanes were ducking in and out of smoke plumes from factories and power plants to measure and characterize pollutants; efforts were being made to compare levels of pollutants in and below clouds; still other activities were dense with language like "a fast-response ozone detector for direct flux measurements." But Vaughan and his team had been waiting impatiently for most of July for just this sort of weather system to materialize. It would be the big event of the study. There had been a couple of near misses—high-pressure systems that would move in from the north and west, only to skirt the study area—but for the past twenty-four hours, excitement had been mounting. The weather station had tracked this new system as it appeared over Canada, crept southward over the Upper and Lower Peninsulas of Michigan, skirted a corner of Indiana, and finally settled over central Ohio. Now the team was on red alert.

Hours before my visit to Mission Control, members of the team had released a "tetroon"—a four-sided, inverted-pyramid-shaped weather balloon—near the high smokestack of the Conesville Electric power plant, a facility that, not incidentally, was the sixteenth-largest emitter of sulfur dioxide in the eastern United States. Released directly into the smokestack's emission plume, the tetroon's sole purpose was to float freely along with the power plant's plume as the plume flowed into, and then was engulfed by, the parcel of high-pressure air.

As the stagnant weather system's winds had swirled slowly, the tetroon had moved along perfectly, all the while beaming back its location to an FAA air-traffic-control center near Cleveland. At Mission Control, a new pushpin was stuck into the map periodically to indicate the tetroon's latest location, So far, it had spun in a slow, spiraling, clockwise loop—an "anticyclonic" motion typical of winds around a high-pressure system.

A PEPE tends to occur when a high-pressure system's usual, southeastward progress is suddenly stalled. This particular PEPE was not destined to rank among the most severe pollution episodes on record; those tend to occur when the movement of a high-pressure cell is retarded by a major, turbulent weather system—such as a hurricane—prevailing off the East Coast of the U.S. While the high is stalled over a heavily industrialized region, it inhales thousands of tons of hydrocarbons, carbon monoxide, carbon dioxide, microscopic particles of ash and soot, sulfur dioxide, the oxides of nitrogen, and trace amounts of toxic metals such as cadmium and mercury. Once those thousands of tons of gases and solids from hundreds of individual sources have been gathered into the system, they are swirled and blended, reacting with one another, and are cooked by the energy of the sun, to form wholly new chemicals, including acids. Researchers know that such elaborate chemical transformations occur, but little is known about precisely how they take place. So, for the duration of this PEPE the team would dispatch a fleet of flying and mobile ground laboratories to learn a bit more about atmospheric chemistry and air pollution. Bill Vaughan had arranged for me to board one of those aircraft for a five-hour mission into the very heart of the PEPE, the atmospheric cauldron where acid rain is made.

James Foley worked himself around the small computer, the chart recorder, the piles of printout paper on the floor, and the astonishing array of electronic equipment mounted into steel frames against the Cessna's bulkhead. He squeezed and snaked his way forward past those hundreds of pounds of hardware until he'd positioned himself in a crouch just behind and between John Mitchell, the pilot, and me, in the copilot's seat.

The twin-engine Cessna normally seated a half-dozen passengers comfortably; now all but one passenger seat had been removed, and the plane, dubbed *Chem One* for the duration of this mission, had been stuffed with monitoring gear like a holiday turkey. The cabin was cramped, and overheated from the labors of all the electronic gadgets. Lights glowed and blinked; a stylus hopped on the chart recorder; needles trembled on the meters. Student assistant Mark Freiberg, wearing only a pair of gym shorts because of the oppressive heat, lay prone and marginally airsick in the undulating tail section of the aircraft. Foley normally had the luxury of a cramped seat behind a computer, but now was forced to squat up forward between Mitchell and me.

Foley asked for the microphone.

"*Chem One* to *Relay*."

"*Relay*," a tinny voice acknowledged from the plane's loudspeaker.

"*Relay*," said Foley, "advise Mission Control there's no airflow registering on the S-O-Two Melloy."

Relay was another small airplane, circling somewhere between us and Columbus, forwarding our messages, since we were now too far north to contact Mission Control directly. We listened as the tinny voice forwarded the message, paused, then came snapping back to tell Foley which knobs and switches to fiddle with to bring the malfunctioning sulfur-dioxide monitor to life. Foley struggled back to the troublesome monitor, performed the requisite adjustments, and, exasperated and perspiring, struggled forward again. Still no reading, he told *Relay*, and another series of messages ensued before the monitor suddenly began to behave, apparently of its own volition.

We were flying through a sea of dense summer haze about four thousand feet above a green patchwork of central Ohio farmland, and we were heading for the last reported location of that signaling tetroon as it floated along in this PEPE. Our goal was to fly to a position near the tetroon, then execute a series of patterns around it. First, we'd fly a square forty miles north, east, south, and west, then we'd repeat the same grid at a higher altitude. Next we'd perform a tight, corkscrew pattern, ascending to about six thousand feet, followed by a descending corkscrew to within a few hundred feet of the earth.

All the while, samples of the air around us would rush hissing into the cabin through a series of air-intake pipes mounted on the outside of the fuselage, pouring through a network of hoses into the complex of pollution monitors, to be electronically tasted, sniffed, and analyzed. The instruments would measure levels of ozone, respirable particulates, sulfur dioxide, oxides of nitrogen, sulfates, hydrocarbons, trace metals, and more, all in an attempt to gather more data about the enormously complex chemical reactions that regularly occur thousands of feet above the earth's surface.

At the airport, just before climbing into *Chem One*, I had looked up and seen a cloudless blue sky. It was a hot, sunny day, perfect beach weather. But although the sky was blue, there was that light, gauzy haze so typical of summer skies in the eastern U.S. If I hadn't been heading skyward to take a look at that haze, I might not have given it a second thought. But that haze is far from a natural phenomenon. In fact, it is largely a chemical soup made up of microscopic particles of air pollution and water vapor that has collected on those particles.

After we completed our first series of grid patterns and began to ascend in a tight spiral up through the haze, the characteristics of this pollutant soup became more obvious. The haze thickened, turning to a dense, translucent brown as we rose, until at about six thousand feet we could have been flying through a bowl of beef broth.

"Sometimes," said John Mitchell, "it gets so bad you can't even see over the nose of the plane to the ground. Then a cold front will move in and clear it all out."

Suddenly we rose out of that thick haze altogether. Above us, the sky was a brilliant, clear azure; below, that layer of brown pollutant crud was spread out flat as a tabletop as far as we could see. The effect was that of peering downward through an immense sheet of brown smoked glass, with the patches of Ohio farmland muted, almost colorless, beneath it.

Foley, back in his seat now, looked up from the aviator's map and the plastic navigation plotter in his lap. He winced as we spiraled back down through the top of the haze layer and into the PEPE itself.

"Just think," he said. "You have to breathe that shit all the time.

James Foley was rightly concerned. That the various unpleasant chemical components of a highly polluted air mass such as this PEPE regularly pass in and out of the lungs of humans is beyond dispute. Most recently, a report by Congress's Office of Technology Assessment stated that as many as 51,000 people may have died prematurely in 1980 from illnesses caused or aggravated by air pollution, particularly from the sulfur pollution involved in the production of acid rain. By the end of the century the death toll could climb to 57,000 a year in North America if emissions continue at current levels. These may be conservative estimates. Researchers Robert Mendelsohn of the University of Washington and Guy Orcutt of Yale University sifted through more than two million death certificates and census figures from some three thousand counties in the eastern United States, and fed that data into a computer for comparison with EPA records of sulfate air pollution from the same areas. The two scientists found a clear correlation between increased sulfate levels and premature death, particularly heart disease. Based on their data, they estimate 187,686 premature deaths per year in the United States.

These studies do not prove conclusively that pollution has caused the deaths, since no researcher yet has been able to follow a mote of air pollution as it leaves a smokestack, travels miles through the air, enters a human lung, and contributes to a long, slow process that could lead to cancer or emphysema that manifests itself years later.

But there is one incident of severe human devastation from air pollution that cannot be denied. On December 4, 1952, a new word was coined in London: "smog," a combination of "smoke" and "fog." Another term was coined at about the same time: "killer fog."

On that day in December, a large, cold parcel of air moved over the city of London, where coal was burned not only to produce electricity but also for residential heating. The chill on that December day was a bit severe, but nothing extraordinary. Londoners stoked their coal stoves and went about their business. Those who bothered to look skyward surely noticed that although the day was cloudless, an increasingly dense haze was forming over the city. Smoke rose in ramrod-straight columns from residential and industrial chimneys, then, suddenly, just a few hundred feet overhead, stopped rising altogether. At that point, the smoke turned abruptly to the horizontal, flattening out as if it had encountered an invisible lid. Pilots flying in and out of the city that day must have noticed the same phenomenon we were to witness years later above the PEPE: the haze over London had a distinct, flat top a few hundred feet above the earth. The air above the haze seemed extraordinarily clean; the air below was an ugly, dark fog.

London was experiencing a temperature inversion. Usually, as altitude increases, air pressure decreases, and as pressure decreases, so does temperature. Thus, the air tends to be warmer nearest the earth and cooler as it rises, since the various molecules that make up the atmosphere are more active under higher pressure.

An inversion reverses the usual order of things: a layer of warm air forms over a layer of cooler air. On a small scale, inversions occur routinely on calm, clear nights when the surface of the earth cools the air directly adjacent to it, particularly in low-lying areas, such as small valleys around streams and ponds. When this sort of inversion occurs, fog and mist often form.

The inversion that settled over London in 1952 was of a different sort. A so-called subsidence inversion, it had been brought on primarily by a high-pressure air mass that had stalled over the city. The influence of such air cells can be vast, covering thousands of square miles. The center of the cell is like the center of a merry-go-round. Winds around the outside flow in a clockwise, anticyclonic direction (counterclockwise in the southern hemisphere), spinning outward toward the nearest low-pressure system, which is swirling in a counterclockwise, cyclonic direction. (This, too, is reversed in the southern hemisphere.) The greater the difference in pressure between the center of the high (which in the United States might be located, say, over Ohio) and the low (which might be over Virginia), the more forceful the flow of wind from high to low. These stronger winds tend to mix and blend warm air with cold, so that the usual condition of steadily decreasing temperature with increase in altitude prevails.

At the very center of the high, however, winds are usually nearly absent, thus allowing air to sink to lower altitudes without being constantly churned back upward by normal turbulence. And, as the air sinks, an increase in pressure causes it to heat up. Eventually, the relatively warmer air will have piled into a veritable mound over a shallow layer of cooler air. This is an especially stable, and stagnant, condition, an inversion that effectively clamps a lid over a piece of the earth.

Consider what might happen to a hot-air balloon released into such a system. The balloon flies because warm air tends to rise, and the balloon will continue to ascend as long as the air inside it is warmer than the air outside. But should the balloon reach a level where its own air is precisely the same temperature as the air outside, the ascent will stop.

This is precisely what happens to pollutant particles and gases emitted under the tight lid of a temperature inversion.

As Londoners stoked their stoves, operated their factories, and drove their automobiles that day in December, warm, pollutant plumes rose to the ceiling of warm air. Since there was no horizontal wind to carry them away, these pollutants began to collect under the lid, the various plumes crisscrossing and blending, the hydrocarbons and oxides of nitrogen and sulfur dioxide and particles of soot and poisonous trace metals interacting with one another to form new combinations of pollutant chemicals, just as new chemical compounds were forming years later in our episode over Ohio.

But there was a profound difference between the two episodes. Luckily, ours would move on within a day, and its ceiling would remain several thousand feet high. London's PEPE pressed itself tight against the city and remained for days. The total concentration of pollutants, in those days before *any* meaningful pollution control, was tremendous.

By December 5, the pollution over London was so dense that the city remained in an extended, hazy twilight. Two days later, the inversion was still clamped tight over the city, and hospitals were overflowing with residents who had developed severe respiratory and cardiovascular ailments. Finally, four days after it began, the great killer fog of London was blown away by another intruding weather system, but those days in December were to go on record as the greatest single air-pollution disaster in history. Uncounted thousands were temporarily incapacitated by the incident and British authorities calculated that four thousand people died as a *direct* result during and shortly after the tragic episode.

It would be impossible to trace the precise chemical mechanisms that killed any one of those victims, but it is almost certain that one of the primary causes was sulfur-dioxide gas emitted by those residential and industrial coal-burning stoves and furnaces. By itself, sulfur dioxide in its gaseous form is a lung irritant, but it now appears that something else, much more ominous, was happening.

If a molecule of sulfur dioxide (SO_2) can somehow bond itself to an additional single atom of oxygen and then to a molecule of water, the result is a molecule with two atoms of hydrogen, one of sulfur, and four of oxygen: H_2SO_4, sulfuric acid.

There is no longer any question that such transformations can—and do—occur in a PEPE and, to a lesser degree, in an air mass with even less concentrated pollution. There is no longer any question that acid forms regularly, constantly, in the atmosphere over the industrialized world and is brought to earth in the form of precipitation. And there is no longer any question that its damage is *not* limited to occasional, freakish incidents in which human lives are lost, but rather to an alarming range of ecological destruction.

The earth's atmosphere can best be described as an ocean—an ocean of air that helps to nourish life. Like the oceans, the atmosphere is characterized by currents flowing and swirling both laterally and vertically—movements that cause airborne substances to mix in much the same way that many substances do if poured into water.

But even though the atmosphere is a mixture of hundreds of different chemicals, two gases, nitrogen and oxygen, together constitute more than 99.9 percent of the molecules in the air. Beyond these two ubiquitous gases, there are hundreds of "trace" chemicals—elements and compounds that can be measured only as parts per million or even parts per billion of all the molecules present. The various trace species in a clean, unpolluted atmosphere include a variety of carbon compounds, methane, formaldehyde, the oxides of nitrogen, ammonia, ozone, sulfur dioxide, hydrogen sulfide, dimethyl sulfide, helium, neon, argon, krypton, xenon, and, of course, both carbon dioxide and water. (Water is present in gaseous, liquid, and solid form.)

Many of these chemicals play vital roles in the nourishment of plant life, and consequently, they *also* play a vital role in animal life, since animals obtain their nutrients directly or indirectly from plants. Consider carbon and nitrogen. Both constantly move in great cycles out of the atmosphere into the living tissues of plants and on into the complex webs of consumption and decay that constitute the intricate weave of life. Plants use carbon dioxide to build new cell matter through the photosynthetic process: they take in the gas through their leaves and then use carbon to help synthesize complex organic molecules of sugar and starch. The sugar and starch, in turn, serve as storehouses for the sun's energy. Leftover oxygen is cycled back into the atmosphere. Similarly, plants use nitrogen as the fundamental building block of protoplasm, the cellular stuff of life itself. Animals (including humans) obtain plant protoplasm in the form of proteins either by eating plants or by consuming other animals that have eaten plants.

Many elements are continually being cycled and recycled into and out of the atmosphere. They fall out of or are washed out of the skies and into the soils, lakes, rivers, and oceans of the earth. They are borrowed for a time, to be used as the building blocks of life, then eventually cycled back into the atmosphere. The loop continues, and continues.

What is most remarkable about all of this is that because of this great atmospheric engine, a handful of seeds scattered on a bare patch of soil can become a forest literally made out of "thin air." That is, a towering white-pine forest—thousands of tons of lumber, bark, branches, and needles—can assemble itself by seizing and using chemicals in the soil and the atmosphere.

Scientists now refer to the everlasting atmosphere-biosphere cycles of chemicals as the "biogeochemical circulation processes" of the earth. In an unpolluted environment, the whole process is not only beneficial but also vital. Terrestrial plants, the soil, and the earth's lakes and oceans *all* receive these beneficial chemicals as part of the process now known as "atmospheric deposition." But such deposition can become disruptive when the mix of atmospheric chemicals loses its natural balance. Burn a forest to ashes and within weeks the first grasses and herbs will begin poking through the devastation, and then the entire forest will proceed to regenerate itself in a predictable pattern of plant succession: herbs, shrubs, "pioneer" tree species (birch, aspen), and finally "climax" mature-forest species (maple, white pine). But this same highly adaptable living system simply is not designed to cope with massive injections of man-made chemicals that spew as waste effluent out of factory chimneys.

Since the atmosphere is an almost perfect recycling system, constantly taking up new matter and redepositing it, the amount of sulfur and nitrogen compounds pumped into the atmosphere by both natural and man-made sources must be related directly to the amount that washes or falls back to earth.

Worldwide, humans and their factories and vehicles spew some 100 million metric tons of sulfur dioxide and 35 million tons of nitrogen oxides into the air each year. Together, the United States and Canada contribute roughly 35 million metric tons—that is, 35 percent—of the world's sulfur dioxide. These two types of pollutant gas—sulfur dioxide and the oxides of nitrogen—are the cause of acid rain, with sulfur dioxide by far the greater villain.

But how do those acids form? No one knows precisely. Even scientists who have studied the acid-rain phenomenon for years understand little about the process that can transform gases to acids or the processes that can move those acids hundreds, even thousands of miles from the original source of emission.

Thus the importance of studies like the one that sent *Chem One* into the PEPE to inhale air samples. For the unanswered scientific questions pose the greatest political problem for environmentalists who want acid rain controlled—and fast. Polluters, who point to the enormous costs of further clean-up of sulfur and nitrogen oxides, firmly insist that little should be done to control acid rain until the phenomenon is fully understood. In 1980, William Poundstone, vice-president of Consolidation Coal Company, told the Senate Committee on Energy and Natural Resources, "It would be unwarranted, unjustified, and unwise for the nation to embark on a course of regulatory controls based on scant, conflicting, and inconclusive data."

In his own way, atmospheric chemist Kenneth Demerjian, who was involved in the Ohio PEPE study, agreed with Poundstone's assessment: "The data base really is not very adequate, and if the environmentalists push [for an acid-rain regulatory program] right now, I'm afraid the industries are going to clobber them."

The industrial polluters have a point. The chemical interactions in a weather system like our PEPE are complex, and only by defining these poorly understood chemical reactions can scientists ever hope to establish clear links between sources of air pollution and damage to sensitive ecosystems. If scientists *can* learn everything there is to know about those reactions, it might be possible to design a pollution-control program that will reduce the threat of acid rain and still cost the polluters as little as possible—one that might require them to remove only 47 percent of their sulfur emissions, for example, instead of 50 or 60 percent.

But that, insists acid-rain research pioneer Eville Gorham,

is begging the question. "It's the standard argument," he told me. "We already know that sulfur and nitrogen oxides are causing real damage. I think there's no doubt about that. They say we need more studies, but if we wait for the last *i* to be dotted and the last *t* to be crossed, more lakes will be lost. And they'll *still* be saying it."

The prestigious, and normally very cautious, National Academy of Sciences appears to agree with Gorham's assessment. In 1981, the Academy's National Research Council issued a strongly worded report, "Atmosphere-Biosphere Interactions: Toward a Better Understanding of the Ecological Consequences of Fossil Fuel Combustion," which stated, "Although claims have been made that direct evidence linking power-plant emissions to the production of acid rain is inconclusive . . . we find the circumstantial evidence for their role overwhelming. . . . There is little probability that some factor other than emissions of sulfur and nitrogen oxides is responsible for acid rain." The report went on to recommend a 50 percent reduction in sulfur emissions, as well as sharp cuts in nitrogen-oxide emissions. "The Committee believes that continued emissions of sulfur and nitrogen oxides at current or accelerated rates, in the face of clear evidence of serious hazard to human health and to the biosphere, will be extremely risky from a long-term economic standpoint as well as from the standpoint of biosphere protection." The report added, "At current rates of emission of sulfur and nitrogen oxides, the number of affected lakes can be expected to more than double by 1990, and to include larger and deeper lakes."

No one would openly deny the value of the sort of research that *Chem One* was conducting. Its perspective—the middle of an atmospheric caldron where uncountable chemical reactions are occurring—also offered an ideal opportunity to look at acid rain itself, at what scientists know about how it is made, and at how it can travel astonishing distances before finally falling to earth.

One of the great deficiencies in acid-rain research has been the lack of long-standing, comprehensive historical records detailing the chemistry of rain and snow in North America. But by the 1980s, that deficiency had been corrected. Hundreds of sophisticated rain and snow collectors (as well as a few less sophisticated collectors that look for all the world like plastic mop buckets) are now scattered over the continent. In the U.S. and Canada, over 140 collectors operated by universities and government and private research entities are part of formal, long-term programs for collecting and monitoring precipitation. Most collectors are of a similar design: shaped like a space capsule, with a wide base and a narrow top, a tiny, electrically charged contact projecting from the gadget. When a drop of water or a flake of snow falls onto the contact, the lid silently opens, and for the duration of that rainfall or snowfall, the container collects. (Other types of collectors catch *all* falling debris—rain, snow, dust, twigs, insects, leaves. Still others, most notably an ingenious cloud-water sampler positioned high on Whiteface Mountain in the Adirondacks, collect samples of water directly from clouds that form over the high peaks of mountains.)

To comprehend what these collectors are indicating about the state of the rain, it is first necessary to understand the pH scale, the commonly used yardstick for measuring the acidity or alkalinity of a substance.

On the pH scale, the number 7 indicates a perfectly neutral substance—that is, a substance that is neither alkaline nor acidic. Numbers greater than 7 indicate increasingly alkaline substances. Thus, a very weak alkaline substance such as baking soda might measure around pH 8 on the scale; a more strongly alkaline substance such as household ammonia might measure close to pH 12.

Numbers lower than pH 7 indicate increasingly acidic substances. Thus, the lower the pH, the greater the acidity. A fairly mild acid such as tomato juice might measure pH 4.3; a somewhat stronger acid such as household vinegar, around pH 2.8; a very strong acid such as battery acid, close to pH 1. It is also important to understand that the scale is logarithmic. Each whole-number increase or decrease on the pH scale indicates a tenfold increase or decrease in acidity. Thus, a pH 5 solution is ten times more acidic than a pH 6; a pH 4 solution is 100 times more acidic than a pH 6.

It might seem reasonable that rain would be a neutral substance in nature, but this is not the case. Virtually all rain is slightly acidic under perfectly natural circumstances, because carbon dioxide in the atmosphere combines with water to form weak carbonic acid. This natural reaction normally tends to acidify healthy rain to a slightly acidic pH 5.6 to pH 5.7. There is some dispute among researchers over the accuracy of pH 5.6 as a proper value for "normal" rain. The figure was arrived at by "bench calculation" in the laboratory based on the calculated value of carbon dioxide in equilibrium with atmospheric water. But lately, scientists have been sampling rain in remote parts of the world to try to estimate precipitation acidity before the dawn of the industrial revolution, and have found values more acidic than "normal," suggesting to some that the truly natural value may be a bit lower. Further, researchers have acknowledged all along that natural rain is in some places *less* acidic than pH 5.6 because of the neutralizing effects of alkaline dust kicked into the atmosphere by such activities as viniculture, farming, and construction. Indeed, ice-core samples from glaciers in Greenland indicate that snow pH 180 years ago ranged from pH 6 to pH 7.6. However, there are also some natural sources of chemicals that might tend to acidify rain slightly in some areas. Volcanoes pump out sulfur dioxide and other sulfur compounds, and certain natural, organic decomposition processes create nitric oxide.

But today the rain and snow falling over most of eastern North America, as well as over localized portions of the West, ranges from 10 to 100 times the mild acidity of "normal" pH 5.6 rain. According to the EPA, the acidity of rainfall over the eastern portion of North America now averages about pH 4.5.

And this is only an average. Meteorologist Ray Falconer, who studies rain chemistry from a pleasant log-cabin laboratory operated by the State University of New York on the side of Whiteface Mountain, told me that at least once every year he collects a sample of pH 2.8 rain from those clouds and rainwater samplers: rain as acid as table vinegar.

The lowest rain pH value ever recorded by operators of the extensive European precipitation-sampling network came from a rainstorm over Pitlochry, Scotland, on April 10, 1974, where rain was tested at a pH of 2.4. Values of 2.7 were reported in Norway the same month. In the United States, it appears that one long rainfall in the autumn of 1978 over Wheeling, West Virginia, holds this dubious record: values under pH 2 were recorded. As a point of reference, this is a level six to eight times more acidic than vinegar, and some five thousand times the acidity of normal rain.

Today, the area receiving highly acidic precipitation (pH 4.6 or lower) in North America includes large parts of eastern Ontario and Quebec, most of Newfoundland and Nova Scotia, and large parts (or all) of the states of Wisconsin, Minnesota, Michigan, New York, New Hampshire, Vermont, Maine, Massachusetts, Connecticut, Rhode Island, Iowa, Missouri, Illinois, Indiana, Ohio, Pennsylvania, New Jersey, Delaware, Maryland, Kentucky, Virginia, Tennessee, North and South Carolina, Arkansas, Mississippi, Alabama, and Georgia.

But even though acid rain is falling over an astonishingly

wide area, not all regions are affected equally. There is little question that acid rain is destroying human-made structures all over the world; its effect on limestone and marble is particularly profound. In Athens, authorities finally removed a series of statues on the Parthenon whose features had been wiped out by acids and other air pollutants. In Washington, D.C., National Park Service guides explain that the huge, conical stalactites that have formed in the chambers beneath the Lincoln Memorial are the result of acid rain eating away the marble. Further, there is little question that the air pollutants associated with acid rain are causing early deaths and morbidity. But aside from acid rain's damaging effect on monuments and other stone structures and the effects of related pollutants on the human respiratory system, the clear ecological damage that has resulted from acid rain in North America is quite strictly limited to a few relatively small regions in the United States and a number of much larger regions in eastern Canada. It is perhaps the most bitter of ironies that industrial states like Ohio and Illinois, where the bulk of the acid-rain problem appears to originate, are often immune from the most obvious ravages and that some of the most remote wilderness lakes—some of the continent's greatest natural jewels—are distinctly threatened, or already destroyed, even though they lie hundreds of miles from the sources of their destruction.

Why the disparity? Much of the earth's land and water contains a substantial reserve of alkaline minerals, which can effectively neutralize onslaughts of acidity. Those areas that through accidents of geology have little such neutralizing capacity are prime candidates for acidification.

Surely the stickiest of political problems surrounding acid rain is its ability to move astonishing distances, across state borders (from, say, Illinois to upstate New York) and across international boundaries (England, Germany, and the Eastern

Bloc are polluting Scandinavia). In fact, its long-distance movement is directly related to sulfur dioxide's efficiency in converting to acid rain. It appears that sulfur dioxide can remain in the atmosphere for up to five days—and thus move as far in those days as a weather system can carry it. The longer the gas can remain aloft, the more it can react with other chemicals to form acids. Although scientists, at this writing, still do not understand precisely how the transformation process takes place, evidence seems to point increasingly to chemical agents, such as hydrogen peroxide, in the atmosphere that act as a chemical-processing factory inside a cloud droplet to combine sulfur dioxide with oxygen. Once sulfur dioxide is oxidized, it dissolves readily in water to form sulfuric acid. Further, air movement within a polluted weather system, the amount of sunlight, and levels of temperature and humidity all seem to be factors in the rate of conversion of pollutant gases to acids.

If it is difficult to conceive of huge, regional masses of hazy, polluted air moving such enormous distances, perhaps the vantage point of Walt Lyons will help make the picture clearer. Lyons is the weatherman for a television station in Minneapolis, and with his unique brand of frenetic energy, he delivers the weekend weather forecast to that city's viewers. Few of those viewers are aware that Lyons is also president of the meteorological consulting firm Mesomet, Inc.; that he holds a Ph.D. in geophysical sciences; and that he is widely regarded as a leading authority on the relationship between meteorology and air pollution. He and his firm were, in fact, participants in the same study that had sent *Chem One* flying over Ohio. But Lyons's vehicle was of a different sort; namely, a weather satellite circling hundreds of miles above the planet and beaming information back to a computer system at the University of Wisconsin.

On satellite photos interpreted with the aid of a computer,

Lyons was able to observe a hazy blob, like our PEPE, as it circled lazily for days over the central United States, inhaling more and more pollutants. As the blob swirled, colorless, invisible gases were transformed into submicroscopic particles—particles smaller than viruses but light-refractive enough to create a heavy haze, lowering visibility from ten miles to five miles to three, until the whole soupy weather system finally was nudged away. On his satellite photos, Lyons tracked the clockwise-swirling winds around the cell of high pressure at the center of the blob as they moved out of the Ohio River Valley, west toward Missouri, then north over Iowa, Minnesota, and Wisconsin, where the system stalled once again.

Lyons says, "One of the strangest things we've seen on these satellites pictures is these large masses of clouds with lots of little holes punched in them. For a long time, we couldn't figure out why there were these holes, but then we realized that what had happened was rain: the smog had been absorbed into the raindrops and fallen out of the blob. We've verified this by comparing precipitation data taken from the ground with the satellite pictures of these holes. What we're seeing is acid rain actually being made."

Not only that. The Swiss-cheese-like holes Lyons was seeing—those punch-outs of clear air in the middle of a hazy, polluted blob—appeared directly over northern Minnesota and northern Wisconsin, where there is scant buffering capacity to cope with acids in the soils and waters, where there is much to be lost if the chemical makeup of such weather systems cannot, somehow, be changed.

3

There are four links in the acid-rain chain, and all four must be present before it becomes an ecological problem. First, pollutant gas must be available; second, an atmospheric kettle that can transform gas into acid is necessary, as is a weather system that can transport the acid hundreds of miles; the fourth is what scientists call a "sensitive receptor," a piece of the natural world that is susceptible to acid damage. In the case of lakes, not all waters are threatened equally. Whole truck tanks of strong sulfuric acid could be poured into many lakes in the United States without causing long-term acidification.

But stand at the shoreline of any of the acidified lakes in Sweden, Norway, Canada, or the northeastern United States and what is most striking is that the waters themselves, and the forests surrounding them, all look so similar. The lakes are cold and clean; the forests are coniferous. It is no coincidence, for when biological death comes, it comes to waters that have a geological common denominator.

At my springtime campsite in the canoe country, we could literally sit upon this common denominator: the smooth shelf of granitic bedrock that juts into the lake. This slab is part of one of the oldest exposed landmasses on earth, part of a vast piece of geologic structure that had its origins in a series of upheavals, fiery eruptions, and floods over a period of at least 2.7 billion years. During that vast sweep of geological time, the canoe country was submerged under ocean several times, and later, there were volcanic eruptions, spewing and blasting flows of lava and molten rock. The earth's crust tilted and shifted, throwing up Himalaya-like mountain ranges, which were slowly eroded to nearly flat in ensuing aeons. The forces that shaped the present-day canoe country were themselves born one and a half million years ago: perhaps the earth's axis wobbled, or for some reason, the sun's energy cooled slightly, and the Pleistocene glacial epoch began. It took a decrease of only five degrees in the earth's average temperature to prevent snow from completely melting during summer in the far north. Decade after decade, then century after century, millennium after millennium, snow piled upon snow, creating pressures of up to seven tons per square foot, first jamming snow crystals together to form ice, then forcing the ice crystals themselves to begin disassembling and flowing forward and outward like liquid. Meanwhile, the vast glacier began to serve as a giant air conditioner, cooling the north winds and further lowering the earth's temperature until the glacier finally had extended itself southward across most of what is now Canada and well down into the northern tier of American states. Four times (at least), the glacier moved into the formerly temperate regions of the hemisphere; four times it retreated.

The last lobe of mile-high glacier retreated from our slab of granite roughly ten thousand years ago. The Swiss naturalist Louis Agassiz, who in the nineteenth century first pieced together the story of the Ice Age, called the glacial sheets

"God's Big Plow." Across eastern Canada, northern Minnesota, northern Wisconsin, northern Michigan, northern New England, the glacier left behind a vast, exposed, solid plain of scoured bedrock: the Canadian Shield, two million square miles of barren rock, stretching from Winnipeg to the Atlantic Ocean, the northern United States to Hudson Bay; no animals, no plants, and precious little soil.

But, mote by mote, the soil came back over the centuries. Raindrops flowing across the face of the rock weathered away bits of mineral. Possibly the first life forms to return were the uniquely adapted, symbiotic fungi and algae that together form mossy lichens on barren rock, the fungi tapping the rock for minerals, the algae producing new plant-cell matter via the engine of photosynthesis. And through the decay of the lichens and the dissolution of rock by rain, soil formed, a few thin millimeters at a time. Into the accumulating soil came grasses and herbs, then shrubs and trees. By the time that pancake of soil was two inches thick, the land and waters were explosive with life. Balsam and birch, sweet fern and dewberry, trout and pike, eagle, osprey, beaver, fisher, marten, bear, wolf, moose, laughing, yodeling loons.

Around 7000 B.C. the first humans settled in the by then temperate and comparatively lush region, gathering herbs and berries and hunting with stone-tipped spears. By the time the first Europeans arrived on the Shield, some 300 years years ago, the descendants of those early Paleo-Indians, the forest Sioux and the Ojibwa, had developed the bow and arrow, clay pottery, the bark-covered wigwam, fish hooks and nets, and the ideal modes of transport: toboggan and snowshoe for the winter, and, for the remainder of the year, the true vehicle of the Canadian Shield, the light, tough birchbark canoe.

Much of the Shield remains sparsely settled and wild. Here and there across its millions of square miles are rich deposits of iron ore, copper, nickel, and other minerals, so there are scattered mining regions: Minnesota's Mesabi iron range; Michigan's copper country; nickel at Sudbury, Ontario; copper at Noranda, Quebec. There is little agriculture, but there is much logging, some trapping, and, because the tens of thousands of potholes and trenches dug by the glacier have filled with water and become clear northern lakes, there is substantial summer recreation and tourism.

I was camping in the Boundary Waters Canoe Area with friend Brian Rusche one autumn when he tossed half a cup of water onto the soil beneath his feet and said, "There it goes, on its long trip to Hudson Bay." Indeed, that is precisely where billions of molecules of that half a glass were bound. As the water leached down through the few inches of soil along the bedrock contour of the hill we sat on, some would be trapped and used by the roots of plants along the way. But much of that bit of water would quickly find its way into the lake below us—a lake that drained into lakes flowing into lakes flowing into rivers flowing, at last, into Hudson Bay, five hundred miles to the north. On the Shield, when it rains, the water moves almost unimpeded into the nearest lake or stream. And that is the key.

The thin soils here contain only minimal amounts of alkaline "buffering" compounds, such as limestone, which could neutralize acid as it flows through the soil.

Similarly, there is only limited acid-neutralizing capacity across the whole vast sweep of the Canadian Shield, including portions of northern Minnesota and Wisconsin, the upper peninsula of Michigan, the Adirondack Mountains of New York, much of northern New England, and most of the entire eastern half of Canada, from the Atlantic Ocean to the northern Great Plains. There are other low-alkaline regions, including the Appalachian Mountains in the eastern U.S. and the mountain ranges of the West. In northern Europe, the Fenno-Scandian Shield underlies much of Norway, Sweden, and Finland.

Let's go from our slab of rock by the waterside back into my green tent that spring morning by Rose Lake. Consider one of those little spheres of rainwater that sit on the wall of the tent: two atoms of hydrogen and one of oxygen clamped together by the awesome force of covalent bonding; a simple chemical that is known to exist in but one spot in the universe, the little, green, wet jewel called Earth.

But this simple chemical has astonishing properties. Its energy-storage capacity is enormous. Through capillary action, it can pull itself to the tops of the highest sequoias; it is so powerful that it can sculpt a Grand Canyon, yet it is so gentle it can nourish a blade of grass. Life began in the oceans, and, for all practical purposes, it continues only in the water, for all living cells are completely surrounded by the water that moves so freely in and out of cell walls, bringing nourishment and carrying away waste.

And the simple molecule is one with a unique and, to chemists, a surprising structure: the hydrogen atoms arranged in a Y pattern along the oxygen atom in such a way that the substance is biopolar, both positively and negatively charged, like a bar magnet. This property allows it to intrude between the components of virtually any other inorganic compound, that is, it can dissolve nearly anything. Even a chemist's beaker filled with "pure" water quickly becomes a mild aqueous solution, for it has already dissolved infinitesimal traces of the beaker glass.

Thus, the rains in nature are not pure water, and never have been, for they will dissolve almost any atmospheric solid or gas. During a typical, moderate rain shower, some five million drops of water will fall on each acre of land per second, with a force of 2.3 pounds per square inch per drop. This force grinds up rock and moves soil, forming muddy rivulets and, by dissolving compounds in soils, liberating the calcium, potassium, iron, phosphorus, and other nutrients as this chemical fabric flows over the earth and into its lakes and its rivers. The rivers and lakes will not be stopping places for the water, of course, for it will continue on, evaporating, moving, and falling elsewhere.

Let us construct an imaginary aquarium, into which we place a freshwater fish—say, a brook trout. We'll feed this fish a natural diet of insect larvae and smaller fish and tiny adult insects, and we'll let it swim happily about, while it can. But into this aquarium, drop by drop, we'll add bits of sulfuric acid.

For a time, nothing will happen to the fish. If we have taken our water from a healthy lake or stream, there will be some level of dissolved mineral salts in the water—alklaline substances that, like microscopic antacid tablets, will effectively combine with, transform, and neutralize the acid molecules, rendering them harmless.

Finally, though, the buffering minerals all will be consumed by the acid, and the very nature of the water in our aquarium will suddenly begin an insidious, invisible change. Drop by drop, the water will become more acidic, and the composition of the solutions that move in and out of the fish's cells will also begin to change.

Suddenly, the chemistry of the trout's blood plasma will begin to change. Less calcium will be available for bone tissue, and if it is a young, still-growing fish, its spine will begin to twist toward an *S* shape and its head will flatten. If the fish happens to be female, eggs may develop but perhaps will never be released. If they are released and do become fertilized by a male, the embryos are unlikely to develop fully because of acid contamination of vital enzymes.

If we are truly to duplicate natural conditions, we will introduce some aluminum into the water, just as acid would liberate normally insoluble aluminum from the soil of a watershed into a lake. If our fish has somehow managed to survive, the toxic aluminum in concert with the acid will begin to deteriorate the tissue of the oxygen-giving gill. In an attempt to protect the gill, the fish might secrete a coating of yellow mucus, but soon the mucus itself will become so dense that it will clog the gills.

A century ago, coal miners would carry a canary into a mine to monitor the air quality of their environment. If the level of explosive and toxic methane gas rose, the canary would die. If the canary's song ceased, the miners knew it was time to get out. Perhaps the trout and salmon in acid lakes are serving this function in the north woods. The fish are vanishing; the canary falls silent.

4

Before breakfast on a summer morning at the Adirondack Loj, I clunked down the stairs in hiking boots and stepped outside. It was well below fifty degrees, and a thin mist rose gossamer from Heart Lake. Thick clouds wrapped and curled around the highest of the Adirondack peaks beyond.

The trail to the High Peaks region begins here at the Loj, a rustic facility operated by the nonprofit Adirondack Mountain Club. The trail leads, after a day-long climb, to Lake Colden.

The previous night, under the yellow glow of a table lamp in the lounge, I'd read a story, "At the Sources of the Hudson" by Vincent Engels, in an old issue of *The Conservationist,* a magazine published by the New York Department of Environmental Conservation. Back in the 1930s, Engels and various fishing partners would pack in with fly rod and fry pan to Colden and other alpine lakes in the High Peaks region, and there, under the shadow of Mt. Marcy, Mt. Colden, and Mt. McIntyre, would pass a day or so sweeping long, gentle casts atop deep, cold waters that would support productive fly fishing even during the dog days of summer. These mountain lakes, the sources of the Hudson River, were once chock-full of brook trout that were, as Engels put it, "beautifully colored, deep-bodied, and delicate in flavor," adding, "the finest trout I have eaten anywhere."

He tells of camping overnight near water's edge in a log lean-to, with balsam for a springy bed beneath his sleeping bag; of brookies rising in the mountain waters all day long. He tells of the time he and a friend left Lake Colden—too early, it turned out—to hike the long trail back to a waiting car, only to learn later that the local ranger had had the best fishing of his life that same evening in the same spot. "I was rowing across the lake," the ranger had told him. "Fish were breaking everywhere, but I didn't stop to cast until I saw the size of some of them that were cruising on the surface. The air was full of little smoky wing-flies with a kind of brown body, and the trout were zig-zagging after them, left and right, tails and backfins and sometimes their snouts too out of the water."

There was a special poignancy to Vince Engels's reminiscence. "No one," he said, "who fished Colden and Flowed Land [another lake in the High Peaks] in the 1930s could have believed that brook trout would ever be lost to these waters that originate in the highest springs and are protected from pollution on every side by miles of state-owned wilderness. But the unbelievable has happened."

Engels's article appeared in 1978, not long after the New York Department of Environmental Conservation had finally figured out what was happening to the high-altitude lakes of the Adirondacks, where trout fishing is the stuff of legend. There are still many cold, clear, and generally unsullied Adirondack lakes and streams that remain ideal for production of the scrappy little fish Mainers call "squaretail"; where mayflies still hatch in clouds on a spring morning; where there is still no choking mucus on gills; where there are no deformed spines, no eggs that refuse to hatch.

But these areas don't include Lake Colden, Flowed Land, and the rest of the High Peaks lakes. Not only could Vincent Engels find no productive fishing in Lake Colden, not only could he wait for weeks on end without seeing a single splash of a brook trout, but there is virtually no hope that trout, indeed any sort of fish, could hope to survive without direct injection of enormous amounts of chemicals to counter artificially the state of those waters. In fact, it is said that Lake Colden has now become so poisoned that fat bubbles of garbage-scented gases sometimes rise to the surface as a result of bizarre biochemical processes on the lake floor.

Back in the 1950s, when some anglers began to notice declines in the productivity of some of the region's high-altitude lakes, state conservation officials tried to fix the blame somewhere: pollution runoff caused by cottage development at lakeside, or perhaps excessive beaver activity. No one suspected that the rain itself had become a poison to the very life it should have nourished.

This day, I had made plans to see Bill Marleau, for Marleau could tell me, and show me, what acid rain can do.

The drive to Marleau's took me past the ski jump for the 1980 Winter Olympics and through the town of Lake Placid. The heat had risen perhaps twenty degrees, and the streets were clogged with cars, the sidewalks with kids with skates slung over their shoulders bound for a competition at the indoor Olympic rink.

From Lake Placid, it is a two-and-a-half-hour drive to the village of Big Moose, more than a hundred miles through the hamlets of Saranac Lake, Tupper Lake, Blue Mountain Lake, and Eagle Bay, through forests of aspen and maple, yellow birch and eastern hemlock. One hundred miles, yet at Lake Placid, I was still a good thirty miles from the eastern border of the Adirondack Park, and at Big Moose, I would still be twenty miles from the western border, for the park's six million acres cover about a fifth of the total land area of the state of New York. The park is unique beyond the fact that it is the largest in the United States. Here, within a day's drive of some fifty million people of the Eastern Seaboard megalopolis, is a vast and rugged place where as recently as 1960 explorers were discovering new lakes not marked on any map. Here there are hundreds of resorts, exclusive sporting clubs, private lodges and cabins, small towns, ski resorts, and the usual assortment of tasteless tourist traps. Here also, within the so-called blue line that defines the park boundaries and where development is tightly controlled by the Adirondack Park Agency, are some 2.3 million scattered acres of wilderness country—roughly the same amount as in the Minnesota-Ontario canoe country (though not, as there, unbroken).

The park contains 100,000 acres of forests that have never been touched by the lumberman's ax; 42 mountains taller than 4,000 feet; 2,300 lakes; and more than 30,000 miles of streams—all protected by a "Forever Wild" law passed by a statewide referendum in 1895. The park also lies on a southern extension of the Canadian Shield and directly in the path of polluted weather systems from the industrial states of the Midwest.

As Bill Marleau told me over the telephone, his house is "the last one before the road ends" in the hamlet of Big Moose, not far from Big Moose Lake. There was a huge stack of firewood beside the modest house and, in the driveway, an orange four-wheel-drive with the seal of New York's Department of Environmental Conservation on the door. I found Marleau in his living room, watching a Yankees game.

Except for a brief stint in the navy during World War II, Marleau has lived in the western Adirondacks all his fifty-some years, and has spent most of his adult life as a forest ranger working out of Big Moose. He told me about how when he was a boy a man would bring big brook trout from

Woods Lake down to his father; how there used to be an old fishing lodge at Woods Lake lined inside with heavy building paper upon which fishermen through the years had scribbled the sizes and weights (maybe exaggerating just a tad) of their most spectacular Woods Lake catches; how Marleau, when he got the chance, leased 335 acres of lakeshore property from the International Paper Company, tore down what remained of the old lodge, and, in 1962, built himself a cabin. There was still great fishing in Woods Lake back when that cabin was built.

Marleau talked of spotted salamanders and kingfishers and tree swallows, the beaver and osprey that once had thrived around his lake. Once, right outside the door of his cabin, he could lay out a fly line with a little furry bucktail fly on the end, work it a bit, and—slam!—into the hook would come a brookie: one he caught around 1960 was a veritable football of a fish, weighing three pounds, but only fifteen inches long.

Marleau asked if I wanted to see the lake, and he and I and his mongrel dog Kelly climbed into the four-wheel-drive and made off down a sand paper-company road. He pointed north, off through the aspens and maple. The lake was just over that two-humped mountain, he said, but the approach was circuitous, and it would take us half an hour to wind our way up to the cabin. We bounced and slammed along as the road got progressively rougher, and the shocks and springs seemed not to help much. Kelly sat on my lap, hugging tight against me to maintain some stability.

Twice on the drive, we stopped so Marleau could unlock gates, first that of the paper company, then, later, with the truck facing up a forty-five-degree incline, his own gate to the cabin road, a two-rut track up the mountain. There was new gravel on the road, and Marleau explained that he had made the scientists buy it and spread it. He was willing to let the scientists use his cabin and property around the lake, but if they were going to drive up and down the road, then they were going to help maintain it, and that was that. On either side of us, the dense woods were in the full, moist flush of summer: a backdrop of blue sky beyond a green canopy of birch, beech, hemlock, and sugar maple; the forest floor sun-dappled. Marleau pointed out a white, scaly growth on the trunk of a beech. He said this had been developing on the trunks of many of the beeches around here, and that he'd never seen it until recently. He coaxed the truck up the incline, until suddenly we were at the top, driving across a concave depression in a granite outcrop where water, clear as gin, poured across our path. We pulled up before the log cabin and walked out to the end of the dock.

The sun was glinting from the lake, the sky blue, the forest verdant, lush. The water itself seemed as clean and clear as any I'd ever seen—in fact, *more* so. Marleau, in his wide-brimmed hat and his sunglasses, surveyed the scene.

"The trouble with acid rain," he said, "is that people can't see it. There's nothing about acid rain a fisherman who comes in once a year can see. You *can't* tell him this water is polluted."

Indeed, by all appearances, the lake is as clear and unsullied as any body of water could be. It has not always been so. Before the 1960s, it was murkier, filled with thousands of suspended bits of decaying organic matter and tiny bits of plankton that had served as the living foundation for the healthy ecosystem's complex of interlinked food chains. The bits of suspended algae, along with lake-bottom "benthic" plants, use photosynthesis to convert the sun's energy to plant tissue. Tiny invertebrate animals in turn consume the plant tissue and then are consumed by larger animals—perhaps amphibians or fish—which are consumed by larger animals. All along the way the sun's energy, the fire of life, is passed throughout the food web.

In the late 1960s, as Woods Lake began to clear up, Marleau noticed the first changes in that web. An avid angler, he was no longer able to catch as many fish, and as the years passed, he found he was catching only larger, more mature fish. Entire generations of the trout population seemed to have been mysteriously wiped out. Then he began finding brook trout in schools at the tiny dam at the lake's outlet, apparently struggling to escape from the lake. The state tried planting thousands of new fingerling trout, but none would survive. Then even the old trout were gone. The disappearing fish turned out to be only the first warning of an assault that before long would extinguish whole species of plants and animals in the lake.

"Used to be," Marleau said from behind me on the dock, "in June you couldn't hear yourself talk for all the bullfrogs in the lily pads. Used to be there were lily pads all along that side of the lake."

But today, where he points, there are only a half-dozen lily pads remaining along the lake's south side, and even if we had stayed long enough for them to start their songs, we easily could have heard ourselves talk, because the frogs are gone from Woods Lake, along with the trout, along with the pair of loons that lived here until about 1969, and the salamanders and otters.

Marleau told me he used to be able to find the best fishing on the lake by following the tree swallows. "Wherever the swallows were circling around the water, there was a fly hatch. I haven't seen a mayfly hatch out there in I don't know how many years." Or as many swallows.

To comprehend how Woods Lake could be transformed from clean fresh water into a mild acid bath, let's hop a thousand miles across the Canadian Shield to the lakeland country of northern Wisconsin to watch as a scientist takes a small, test-tube sample of lake water through the acidification process.

We are at the Rhinelander airport in the heart of Wisconsin's summer-tourism industry. Four times a day, a Republic Airlines jet from Green Bay or Oshkosh or Ironwood comes in low over the upper Wisconsin River. The state's Department of Natural Resources (DNR) flies its fire-spotting and fish and game patrol airplanes out of a hangar here, and there are a few private craft moving up and off the runway. But the main activity on this sunny day is a series of brief appearances of a helicopter filled with gadgets and gauges and carrying a pair of technicians from the U.S. Environmental Protection Agency. The technicians are here as part of a research program operating out of the agency's freshwater-research station in Duluth. The pontoon-outfitted helicopter's mission is to collect samples from northern Wisconsin lakes to help determine if, and how quickly, those lakes might become acidified. Primarily, they are measuring how much capacity those lakes have to neutralize acid input.

Technicians carry the samples into a semi-trailer that is parked behind the DNR hangar. Inside that trailer, a cassette player clamped to a twelve-volt battery plays soft rock, and three young EPA chemists hunch over a long, cluttered laboratory counter and fiddle with various instruments of chemical analysis. Directly from the helicopter, chemist Larry Heinis had already used a device called an "interocean probe" to measure pH, conductivity, light transmittance, dissolved oxygen, and other factors. Now in the mobile lab, Craig Sandberg, Lee Anderson, and Joy Rogalla run some of the same tests with more sensitive, finely tuned gadgets. Portions of each sample are then flown to Duluth or other laboratories for quality-control double-checks and to conduct elaborate tests for trace amounts of heavy metals and exotic chemicals.

The most important of all the tests is a standard laboratory procedure called "titration," a process familiar to every chemistry student and used for decades as a precise test for a sample's alkalinity. Rogalla conducts the titration, first adjusting each 100-milliliter sample of lake water to 25 degrees centigrade. Two electronic probes immersed in the sample read its pH on a digital scale, and the titration begins. Drop by precise 50-microliter drop, an electronic auto-burette, which Rogolla has christened "Otto," adds pure sulfuric acid to the sample. The purpose of the titration is to measure the amount of acid that can be absorbed and neutralized by natural alkaline materials in the lake water. For a time, the acid molecules dripping into the lake sample will quickly find their way to compatible alkaline molecules; the two will snap together, and the pH will remain quite stable. The greater the sample's buffering capacity, the more 50-microliter drops of pure acid it can neutralize. Rogolla bends over and fiddles with the thing. The reading remains constant for a few more moments; then, suddenly, she says, "There." All of the alkalinity has been consumed and thenceforth, with each added drop, the acidity of the sample increases.

A massive sort of titration is precisely what is happening to lakes and watersheds over much of the world. Many contain such phenomenal amounts of alkaline buffers that they are virtually immune to the harmful affects of that process. But this is not so for Woods Lake and others located in the Canadian Shield or in similarly low-alkaline terrain. A mild sulfuric- and nitric-acid solution is dripped into the Woods Lake watershed not at the rate of regularly spaced single drops but at a rate of millions of dilute drops per second, with rain or snowstorms, even with mist and fog and, for that matter, with the fallout of tiny, dry, acid-forming sulfate and nitrate particles that can dissolve to form acids. Some is titrated directly into the lake, but about 90 percent of precipitation falls into the terrestrial watershed. Some of this water will flow quickly through thin soils over the skull of bedrock and into the lake. Much will first fall upon, then drip from, the leaves and needles of trees. Natural alkaline salts on the leaves of hardwoods may slightly neutralize the acid, but mild natural mineral acids on the needles of pines and other conifers tend to slightly concentrate the acid, which will eventually work its way into surface waters. Even Woods Lake contained *some* alkalinity, but eventually not enough to neutralize the relentless onslaught of acid rain.

The first effects of acidification at Woods Lake probably were felt in the spring. All winter, atmospheric acids had fallen with the snows. When moisture crystallizes to form snow or ice, acids are excluded from the crystal itself; instead, they are concentrated on its outside. On the ground, a snow pack will age through the winter—a dynamic process of partial melting of the crystals followed by freezing, melting, and refreezing. As the crystals melt and recrystallize, they tend to form larger crystals, with acids concentrating even more on the outside. Geologist Ernest Marshall, who has studied snow-core samples in much the same way he and his colleagues might study rock cores, has found that by spring, the acids tend to concentrate in distinct strata near the top of a snow pack.

The first flush of winter snow melt can thus move enormous amounts of acid into the nearest lake or stream. Norwegian studies have shown that 50 to 80 percent of acid pollutants in snow are removed in the first 30 percent of melt. And this slugload of acid can either be the final, rapid infusion that plunges a lake's pH below the critical level or, as it has in many documented cases, it can even *temporarily* acidify shallow areas in a still-healthy lake that contains some alkalinity (a serious problem for both fish and amphibians if it occurs as eggs are about to hatch). For years, scientists will be studying and recording the precise sequence of ecological events that occur as a lake goes acid, but already, a sketchy outline of the process exists.

Since 1974, the government of Canada has been conducting a tightly controlled study of a small lake (called, simply, Lake 223) in a remote region of far-western Ontario. With tens of thousands of acid-sensitive lakes at stake in the province, environmental officials decided to perform an experimental sacrifice of Lake 223 in the hope that close observation would provide a framework for detecting the early warning signals of damage elsewhere. Year by year, government scientists have added measured doses of sulfuric acid to the lake while simultaneously calculating any further additions from rain and snowfall.

The first year, the artificial addition of acid merely accelerated the "titration" process, removing what little natural alkalinity remained in the lake. Since then, an amount of acid roughly equivalent to the amount falling in precipitation on the Adirondacks and eastern Canada has been poured into the small lake, lowering the pH by about .25 units per year (in other words, more than doubling the acidity).

At the beginning of the experiment, Lake 223's pH stood at 6.6. As it fell, there was a gradual, linear increase in such toxic metals as aluminum. At pH 6, tadpole shrimp began to show signs of stress. Normally, maturing shrimp undergo a series of molts of their shells, but that rate began to slow, and an increasing percentage died before reaching maturity.

As the pH fell below 5.8, the shrimp and fathead minnows—both primary diet for the native lake trout—vanished. At pH 5.6, crayfish, another major trout foodstuff, had difficulty hardening shells after molting and, as the pH fell to near

5, began to vanish from the lake. However, the scientists were surprised to learn that through the range between pH 6.6 and pH 5.05, the lake showed a slight *increase* in the productivity of simple algal plants and a corresponding slight increase in the number of zooplankton (the tiny animals that eat those plants). This, the scientists believe, was probably a result of the acidity's removing natural humic materials that lightly stain the water, thus allowing sunlight to penetrate deeper into the water and accelerating, for a time, the photosynthetic process. All evidence suggests, however, that as pH continues to drop, most of the tiny plants and animals will be poisoned.

Much of the remaining acidification sequence is evident at Woods Lake. Looking down into the water's clarity from the end of Bill Marleau's dock was unsettling. There was forest litter—leaves and twigs—on the lake floor, but, eerily, the litter had not begun to decompose; instead it had been preserved as in a giant pickle jar. The lake bottom was like a museum diorama done up in nonbiodegradable leaves and twigs, for the bacteria that would normally be responsible for digesting and decaying such organic matter had been destroyed by acidity.

Decomposition ordinarily releases vital nutrients like nitrogen and phosphorus from dead plant and animal matter and recycles them back into the ecosystem, but since the acid waters of Woods Lake have destroyed those decomposers, fewer nutrients can be released. (Here lies something of an ironic twist: for years, pollution-control efforts have focused on preventing overfertilization of waterways by nutrients from sewage and other pollutants. To an extreme, acidification has a reverse effect, essentially sterilizing much of the life from the waters.)

As the Woods Lake pH dropped through the range between

6 and 5, the number and diversity of plants and animals began to fall sharply. Beneath the surface of the lake, mayfly larvae began to die, and tadpole shrimp vanished.

As acidification increased, one of the most overt signs may have been sudden reductions in frog and salamander populations. Kathleen Fischer, a biologist with the Canadian Wildlife Service, says that amphibians may be "the wildlife most drastically affected" in early stages of acidification, for they typically lay their eggs in shallow temporary pools generated by recently melted snow. Studies by Cornell biologist Harvey Pough showed that in pH 5 water, more than 80 percent of salamander eggs failed to hatch because of incomplete embryo development, compared to a normal failure rate of about 10 percent. Of salamanders that do survive the hatch, many are likely to develop with twisted backbones owing to calcium deficiencies. "They play an essential role in the food chain," says Pough. "It would cause major effects on birds and mammals if they were to disappear."

Indeed, frogs, toads, and salamanders are vital to both aquatic and terrestrial ecosystems, Amphibians are among the top carnivores in small ponds and streams and are important predators of aquatic insects. In turn, they are sources of high-protein food for birds and mammals.

For fishermen, the first warning of acidification will likely be a pervasive shift in the age profile of the fish population. Mature fish can often survive in the range above pH 6, but acid waters disrupt fish reproductive systems, inhibiting gonadal development, reducing egg production, and lowering egg and sperm survival. Low-pH waters can actually lower the internal pH of the eggs. Sometimes, embryos may never develop enough to be released from the egg, or fry may develop only to be wiped out by a sudden acid surge from a storm. Before long, whole generations of fish are absent.

Another sign for anglers might be the sudden disappearance of some *species* of fish. For example, in a lake with both smallmouth bass and lake trout, the bass might be exterminated by the time the water drops to pH 6, while the more acid-tolerant trout might survive to pH 5.5. Brook trout—along with a few species of hardy, "rough fish," like bullhead and white suckers—appear to be the most acid-tolerant, and although reproduction would be impossible, a few adults might survive in acid water below pH 5; however, by pH 4.5, all fish life would be gone.

The acidity of a lake or stream is only one edge of the sword, however. Cornell University biologist Carl Schofield, who first detailed the destruction of fish life owing to acid rain in the Adirondacks, is convinced that high levels of aluminum are greatly responsible for the destruction of brook-trout populations. As acid water flows through the soils of a watershed, it liberates small amounts of metals found naturally in the soil. Once washed into a lake or stream, aluminum disrupts the salt balance in blood plasma of fish or destroys the gills and clogs them with mucus.

As the fish, insects, and amphibians vanish, so, too, must the predators that depend on them for food. Few research reports have attempted to detail the damage acidification can do to birds and mammals, but one Swedish study found a clear relationship between low-pH lakes and decreased diversity of waterfowl, gull, and loon species. Another suggested that impaired breeding among four bird species—the pied flycatcher, bluethroat, reed bunting, and willow warbler—was caused by aluminum from metal-contaminated aquatic insects consumed by the birds. It is clear that when food is gone, wildlife populations must decline. According to a joint U.S.-Canadian report, the North American birds most sensitive to acidification include the loon, osprey, great blue heron, belted kingfisher, hooded merganser, ring-necked duck, common merganser, common goldeneye, red-breasted merganser, black duck, green-winged teal, mallard, northern pintail, American widgeon, spotted sandpiper, common yellowthroat, bank swallow, myrtle warbler, eastern kingbird, and blackpool warbler. The mammals most sensitive include American mink, river otter, common shrew, and muskrat.

At the time of my visit, Woods Lake had declined to a severely acid state of pH 4.5. Lobbyists and public-relations practitioners for the industries that cause acid rain would object to my characterization of the lake's acid state as "severe." As they might point out, Woods Lake is indeed not much more acidic than a banana and even less acidic than tomato juice. But fish, amphibians, and hundreds of other species are not able to survive in water as acidic as a banana.

Journalists, politicians, and even scientists sometimes refer to such lakes as "dead." Perhaps "mutated" would be a better word. As Eville Gorham has said, "It is a mistake to say these lakes are dead or dying. It is the qualitative shift that is so startling."

Although Woods Lake and others once sustained thousands of plant and animal species, now only an acid-tolerant few remain, in what amounts to a dramatic simplification of a formerly complex ecosystem. Acid-tolerant insects like water boatmen are able to thrive in the absence of competition. Healthy lakes might normally contain about ten species of tiny, planktonic crustaceans in midsummer, but surveys of several acidified Canadian lakes in the La Cloche Mountains of Ontario turned up only a single species.

Under such conditions, the lake's normal plant life is replaced. Gorham told me about having gone skin-diving in a Canadian acid lake choked with acid-tolerant filamentous algae. He said, "It was rather like swimming in a dusty old attic." On the lake floor, sphagnum moss might begin to form dense beds. Anaerobic bacteria, which thrive in the absence of oxygen, begin to decompose organic matter under the carpet of sphagnum, producing as by-products methane, hydrogen sulfide, and other gases. As a result, bubbles of gas will rise to the surface, and the scent of ripe garbage will sometimes waft over a highly acidified lake.

In the soil of the lake's watershed, rain acidity continues to work its chemical devastation. Of any facet of acid rain's effects on ecosystems, least is known about terrestrial effects, but it is already evident that acids dissolve toxic metals such as aluminum and mercury in the soil, either carrying them into the nearest surface waters or making them available for plant uptake. Acids tend to convert mercury to its most toxic form—monomethyl mercury—a bioaccumulative poison that can settle in the tissue of aquatic insects and fish not yet killed by increasing acidity and can thus be passed along the food chain into the tissues of any animals—including humans—that eat the fish. In northern Minnesota, Canada, and Scandinavia, high levels of mercury have been found in the fatty tissues of fish in remote lakes far from any possible source of industrial pollution.

On plant foliage, acids strip away the leaf's waxy protective coating, interfering with transpiration, decreasing photosynthesis, and possibly subjecting the plants to more stress from disease. Pine, aspen, and paper birch are trees known to be particularly sensitive to direct attacks of acid.

Meanwhile, soil decomposers may be destroyed, making available less nutrient foundation for the forest. Acids may kill "nitrogen fixing" bacteria that live on the nodules of certain legumes, as well as freely in the soil. These bacteria serve as the little "factories" that convert gaseous nitrogen from the atmosphere to a water-soluable nitrate, essential for plant life.

Even when sensitive soils are able to neutralize acids, the neutralization process itself uses up and strips away invaluable nutrients such as calcium, magnesium, sodium, and potassium.

More troubling, West German scientist Bernhard Ulrich, a soils expert at the University of Göttingen, has reported that as the buffering minerals are stripped away from the soil, acid begins to mobilize aluminum. Aluminum is present in great amounts in almost all soils, but normally is harmlessly bound to other molecules. Once these links are severed, the aluminum invades the fine, hairlike roots of trees, through which the plants normally extract water from soils. The aluminum ravages the cells of the root endodermis, the inner cellular passageway through which water moves, and the trees begin dying of thirst from the crown down. With the root system weakened, fungi and bacteria invade, and trees have little hope for survival.

By 1982, the West German interior ministry was estimating that some 1.2 million acres of its trees were dying already in regions ranging from the Harz Mountains in the northeast to the Black Forest in the southwest. In Bavaria, about 150,000 acres of forest are now comprised largely of dead trees and another 8 million acres are showing early signs of stress. Just across the border in Czechoslovakia's Ore Mountains, 247,000 acres of forest are dead, and reforestation efforts have failed.

Hubert W. Vogelmann, chairman of the botany department at the University of Vermont, has found similar destruction in an intensive, two-decade study of the vegetation on Camels Hump, a 4,100-foot peak in Vermont's Green Mountains.

Vogelmann wrote of a 1965 visit to the wild, virgin red-spruce and balsam-fir forest on the mountaintop: "The trees were luxuriant, the forest was fragrant, and a walk among the conifers gave one a feeling of serenity—a sense of entering a primeval forest.

"But you wouldn't believe it today," he said in 1983. "The forest is collapsing. It looks like somebody dropped a bomb up there."

In the mid-1960s, Vogelmann's interest in the slope of Camels Hump had nothing to do with pollution, but rather the intensive cataloging of forest communities. Since the mountain's slope rose through successive climatic zones, it served as an ideal laboratory for botanists to study the vegetation shift from the temperate zone dominated by sugar maples at the mountain's base, up through a more northwoods environment characterized by a band of birches, and finally up to the subarctic fir and spruce forest near the top.

Nearly half of the red spruce in the uppermost region are now dead. Some of the dead spruce had survived on the wet, foggy slope for three centuries. Now Vogelmann's luxuriant and primeval forest of the mid-1960s is one of naked, dried-up tree skeletons. In some spots, high winds have mowed down whole swaths of dead forest.

Vogelmann's preliminary findings about the destruction of red spruce were widely reported in the acid-rain research community in the early 1980s. But in July 1983 he revealed that his most recent survey showed startling declines in the other forest communities, including a 25-percent decline in sugar maple—a much more commercially significant tree than red spruce.

"That," he told me, "is what's really important. Quite frankly, I've been surprised by the levels of decline. But it is evident now that the total productivity, the biomass of the entire forest, is going down, down, down." Similar reports have recently come from the Adirondacks, the White Mountains of New Hampshire, the Laurentian Mountains of Quebec, the Appalachian Mountains, and from forests in Switzerland, Yugoslavia, Poland, France, and England. (There are fascinating and ironic implications in all of this. The International Paper Company, which owns the forests around Woods Lake, has been a leading activist in industry lobbying efforts to relax air pollution regulations.)

One of the oddest findings to come out of acid-rain research is that sulfuric and nitric acids from the skies may actually be *good* for some vegetation. The Tennessee Valley Authority (itself one of the world's most prodigious dischargers of sulfur and nitrogen oxides) once indicated in an analysis of an EPA study that millions of dollars of crop yield in some sulfur-deficient agricultural areas could be lost if emissions were cleaned up, unless farmers invested in tilling commercial sulfur and nitrogen fertilizers into the soil. Indeed, for a limited period, the sulfur and nitrogen deposited by polluted rain may even help counter the adverse effects of acidification in *sensitive* areas. Says Arne Tollan of the Norwegian Institute for Water Research, "There appears to be no serious decrease in forest growth because of the increase in nitrogen deposition accompanying acid precipitation. We are finding an increased leaching of calcium and magnesium in the soils [which would almost certainly hamper forest growth eventually] but at present the fertilizing effects of acid rain are dominant."

The notion has been seized by some polluters as a "benefit" of pumping crud into the air. But, says an outraged Norman Glass, senior research scientist at the EPA and coordinator of the agency's research in terrestrial acid rain, "The notion that it is essential or desirable, or even marginally acceptable, to continue supplying sulfur indiscriminately by using polluted air masses instead of fertilizer is just that — fertilizer."

Bill Marleau retrieved from his cabin a live trap with a couple of mice in it, and we walked along the shore of the lake to release them.

If Woods Lake can no longer be Bill Marleau's fishing retreat, at least a team of scientists has found a use for it. Nowadays, the lake has become an acid-rain laboratory for researchers from such institutions as Cornell, Rensselaer, Dartmouth, and the U.S. Geological Survey. Perhaps the loons do not holler at dusk on Woods Lake anymore and the chorus of bullfrogs has gone silent forever, but now, in the quiet of an evening, should a raindrop patter onto an electrically sensitized plate in the forest near Marleau's cabin, it will trigger a signal for a precipitation sampler to open its wide maw to the rains. As Marleau and I tramped along the muddy ground on the lake's west shore, we could see, rising above ferns and shrubs, a white plastic dust collector here, a tubular stainless-steel instrument there. There were also lysimeters for monitoring water as it seeps through soil and water collectors under various tree species to collect precipitation as it drips off branches, leaves, and needles.

Woods Lake is one of three in the western Adirondacks that scientists are studying under grants from the Electric Power Research Institute. EPRI, a cooperative research unit funded by six hundred U.S. utilities, is pouring millions of dollars into a project to evaluate the effects of precipitation on the three neighboring watersheds, largely to determine why Woods Lake has become so acidic while the others, Lake Sagamore and Panther Lake, test at readings of a marginal pH 5.8 and a perfectly neutral pH 7, respectively.

Since 1977, EPRI has spent nearly $15 million on acid-rain studies, five million of which has gone to the three-lakes study. And while the electric utilities, which are the major sulfur polluters in the United States, have a great stake in the results of the research, the consensus in the scientific community seems to be that EPRI makes no attempt to influence its funded researchers, that, as New York atmospheric chemist Peter Coffey put it, "EPRI funds first-rate research."

At all three lakes, progress of rain is traced literally through the entire watershed, from the time it falls in the lake or on land or in the crown of a tree (there are rain collectors in the middle of the lakes and in the treetops) until it flows out of the lakes.

The reason for the varying acidity levels in the lakes appears to be—as most scientists have speculated all along—that each watershed is unique in terms of slope of drainage, amounts of buffers naturally in the soil, total water capacity, and lake depth. As to the state of Woods Lake, Marleau himself seemed not so much angry as resigned. "When I built this place," he said, "I really thought I had something." And he shook his head.

We climbed into the four-wheel-drive, and as we bounced and banged back down the road, he told me about the kingfishers that once sat on every beaver dam on a nearby creek, about the spring hole where, even if the fish weren't biting, you "could crawl in on your belly and just see 'em laying in there."

We passed through his gate and were heading home via a different route when he stopped abruptly on a bridge over a gorgeous little stream called Twitchell Creek, which flows and bubbles over mossy rocks and under alders—the kind of water that gives a trout fisherman heart palpitations. He pointed to the gurgling riffles, the dark eddies, the flat water behind rocks where once he could entice a brookie to inhale his bucktail fly. Then he reached for a plastic box on the dashboard, we climbed out of the truck, and I waited while he scrambled down a short, steep fisherman's path to the waterside. He took from the plastic box a strip of colormetric pH test paper and briefly held it underwater. We compared the suddenly altered color with a scale on the packet; it showed a pH of 4.3. "About like a tomato," said Marleau.

5

The day before Bill Marleau took me to Woods Lake, I had called on Marty Pfeiffer at the New York Department of Enviroinmental Conservation's Ray Brook regional office, which is in the eastern Adirondacks, not far from Lake Placid. Pfeiffer, short, broad-shouldered, and dressed in a white shirt and suspenders, was the aquatic biologist responsible for acid-lake surveys in the Adirondacks. He stacked reports and articles into my arms like cordwood and led me down into a brightly lit room in the building's basement, where two walls were covered with huge maps of the Adirondack Park and surrounding regions. The maps were stuck with hundreds of pushpins: red or yellow pins for acid-threatened lakes, black pins for lakes that had already gone "critically" acid.

Somewhat arbitrarily, the New York surveys place that "critical" mark at an acidity of pH 5. Pfeiffer had so far counted 212 Adirondack lakes and ponds below the pH 5 mark, and an additional 256 lakes with a pH betwen 5 and 6. Of the critically acidified lakes, 180 were known to have once supported brook-trout populations.

It was in the Adirondacks that the first signs of aquatic

damage from acid rain in the United States began to appear, for this region is subject to a sort of triple whammy. First, underlying much of the region is the most acid-sensitive sort of geologic structure, which allows rain and snow to pass quickly through thin, poorly buffered soil into lakes and streams. Second, the Adirondacks lie directly downwind of major air-pollution sources in the industrial heart of both the United States and Canada. And, third, many of the Adirondacks' most isolated and once pristine lakes lie at high altitudes. Scandinavian studies and Pfeiffer's own research show that when acid rain begins to fall over a sensitive region, it is the high-elevation waters that first show signs of damage.

What appears to happen, according to Norwegian zoologists Ivar Muniz and Helge Leivestand, is "the loss of fish populations [starting] in the mountain lakes and . . . gradually spreading downstream towards the coast."

Pfeiffer has found that in the Adirondacks, well over 50 percent of sampled lakes at elevations of more than two thousand feet have gone critically acid.

Together, these elements create a severe acidification scenario. Industrial emissions from as far away as southern Illinois, Indiana, or Ohio become suspended in large, polluted weather systems that slowly move east, picking up more and more pollution along the way. When the air masses reach the western Adirondack Mountains, they are forced to rise, which decreases the air's pressure and cools it. Clouds form, and, particularly on the windward side of the mountains, rain falls.

Pfeiffer pointed to the array of multicolored pins on the map. "Look at the specific vulnerable areas in the Adirondacks that have already been impacted." He indicated a tightly formed band of pushpins running north to south and arranged precisely at the first high-elevation terrain that the westerly winds encounter. "There's this streak, here, going from extreme southern St. Lawrence County, with the worst concentration in northern Herkimer County, sweeping through western Hamilton and into Fulton County. Then we have the High Peaks ponds, which are very high elevation.''

Pfeiffer later gave me a photocopied miniature of the map, on which he had marked acidified lakes with ugly little orange-brown smudges that ran in a band about eighty miles long and twenty-five miles wide in the western reaches of the park, with a few more clustered in the High Peaks. I couldn't quite shake the fanciful image of a crew of malevolent Olympian gods with huge orange grease pencils, blithely smudging out great pieces of the Adirondack terrain. On a computer printout Pfeiffer had given me, evidence of the wholesale destruction was unmistakable. ''Listing of Acidified Adirondack Lakes and Ponds,'' it was called: Ale Pond, Avalanche Lake, Bone Pond, Big Moose Lake (the largest of the acidified lakes), Cat Mountain Pond, Flowed Land, Colden, Gooseneck, Grassy, Hog Pond, Gull Lake, Jockey Bush Lake, Nine Corner Lake, Otter Lake, Oven Lake, Oxbarn Lake, Sly Pond, Trout Lake (Franklin County), Trout Lake (Hamilton County).

At the time of my visit, the biologically crippled Adirondack waters represented only about 4 percent of the park's total water acreage. But 24 percent of that acreage was already approaching the critical point, and another 21 percent had yet to be sampled.

There are enormous complications in Pfeiffer's research. Under natural conditions, the chemistry of lakes—even lakes in close proximity to one another—can vary widely. The state of New York wants a near-airtight case against the rain in order to force neighboring states to regulate more tightly acid-rain-related pollutants, and it has fallen to Marty Pfeiffer to provide much of the evidence. Lately he has been writing case histories of individual lakes, using not only chemical data but also any fishing records he can root out. But historical records are scanty. In past decades, waters that were chemically tested at all tended to be those most accessible and most heavily used for recreation—often the largest and therefore more capable of diluting acid. Further, as industrial polluters regularly point out, even when some historical evidence exists, comparisons are often tenuous. The most widely used instruments for measuring acidity today is the electronic pH meter, a hand-held gadget about the size of a transistor radio, but when the original biological surveys were done here in the 1920s and 1930s, an entirely different procedure, the "Hellige color comparitor," was used. Further, state fisheries employees tramping through the forests and over the mountains did not always conduct their experiments with precise scientific rigor.

Today, as Pfeiffer supervises a detailed study of waters in the region, he is battling that old sloppiness, so he has taken to analyzing water samples both by modern techniques *and* by the old colormetric methods, in the latter case doing his best to duplicate field conditions under which old samplers would have been examined. The result? Most, but not all, of the present-day pH readings show increased levels of acidity. However, a few of the measurements show surprising increases in *alkalinity*.

There's a "gimmick" here, as Pfeiffer says. Most of the waters that show little change or a slight alkaline increase are either large waters or waters around which there has been, since procurement of the old data, extensive lakeshore-cottage development. Development tends to cause erosion of fertilizing organic matter as well as seepage from drain fields and septic systems. So, says Pfeiffer, "At the same time that we've got acidification, we've got something else going on. In our experience, eutrophication [nutrient enrichment of waters] tends to offset acidification for a while. If it were not for that factor, the trend would be even more obvious."

But, he points out, for the smaller, higher-elevation lakes and ponds, the trend is already clear. There are now no fish in many lakes where fish once thrived.

The Adirondack Mountain region may be the area most seriously affected in the United States so far, but the damage in the U.S. is by no means limited to the Adirondacks.

Summertime pH measurements from 1,368 lakes in Maine have been taken since the 1930s, and they show a steady acidity increase from an average of about pH 6.85 in 1937 to about pH 5.95 in 1974—an eightfold increase in acidity. Reproduction among native brook trout has now ceased in all sampled Maine lakes at elevations of more than two thousand feet. Lake sampling in Vermont and New Hampshire has been limited, but already, researchers have found a handful of high-elevation acid lakes in the Green Mountains and the White Mountains. Scientists have recorded a seventeen-year acidification trend in the Pine Barrens of New Jersey, where little acid buffering can take place in the area's sandy soils. In 1978 and 1979, shallow groundwater in the Pine Barrens averaged pH 4.3. In the Great Smoky Mountains National Park in North Carolina, perennial mountain streams are reading as low as pH 4.3, with related increases in toxic aluminum, and researchers have found evidence of depressed populations of aquatic invertebrates and brook trout. Acid mountain streams have been found in West Virginia and Pennsylvania.

Much more troubling is the *potential* for damage. Given the hundreds of thousands of lakes and streams in North America, the millions of surface-water acres, acid damage has made only a tiny nick in the total. But beyond the more than 100,000 waters on the Canadian Shield that are acid-sensitive, there are thousands more in smaller, localized regions in the United States, including areas in *all* of the Southern and Eastern states, the Rocky Mountains, the Pacific Northwest, and California. Already, researchers in Colorado have recorded acid-rain levels in the range of pH 3.6.

How quickly might those tens of thousands of lakes and streams turn acidic? As yet, no one knows, but despite the availability of equipment to control acid rain, it continues to pour into sensitive waters to consume, bit by bit, the remaining alkalinity, thus moving those lakes continually closer to that point of no return where the pH will suddenly begin its rapid decline. In the Adirondacks, some 19 percent of the park's *lakes* had become critically acidified by 1983. However, since small, mountaintop, brook-trout waters like Woods Lake tend to acidify first, this represented only about 4 percent of the park's lake *acreage*. But perhaps Marty Pfeiffer's experience in New York contains the most ominous warning: "I made a prediction about four years ago," he told me before I left. "I said we would lose 25 percent of the acreage of brook-trout waters by the end of a fifteen-year period. Since that prediction was made, we've already discovered a considerably greater total acreage [of acidified waters] than that."

6

On a midsummer morning in a café in Ely, Minnesota, a man eating breakfast told a companion about a nephew who had come north from Minneapolis to do some fishing. The nephew had fished for days but hadn't caught a thing.

"Probably," chortled the man's companion, "he figured it was because of acid rain."

Later the next autumn, in a rustic tavern in the tiny village of Dorset, Ontario, north of Toronto, the owner of a local marina told his beer-drinking companions of a city dweller whose boat was still moored in a local lake. It was the tail end of November, and the small lakes around Dorset were already rimmed with a thin skin of ice. Water temperature was down to the point where, virtually any minute, the chain reaction that leads to sudden winter lockup of the lake could occur, so the marina owner had telephoned the city fellow to warn him that his boat might be damaged if he didn't come up now to pull it out of the water or pay someone local to do it. The city dweller wanted to wait until the following weekend so he could do it himself.

"Maybe," said the marina operator, "he thinks the acid rain will keep it from freezing."

Chuckles.

As the two incidents demonstrate, some residents of the areas most threatened by acid rain remain skeptical about the threat. But what is truly noteworthy is that by 1981, a group of men eating breakfast in a Minnesota café or watching television in an Ontario saloon would even know enough about the phenomenon to make jokes, for if the term "acid rain" had come up in conversation in these places even three years before, chances are the speaker would have been met with blank stares.

Acid rain became suddenly well known in 1979 and 1980. From obscurity, it became a priority concern of such leading environmental organizations as the Sierra Club, the National Audubon Society, Friends of the Earth, the National Wildlife Federation, and the Izaak Walton League; it also became one of the dominant political issues in Canada and the most heated item of dispute between Canada and its neighbor to the south, the United States.

Yet acid rain is neither a new pollutant nor a newly discovered one. In fact, the thread that leads directly to the discovery of acid rain extends back more than a century. (Less directly, fundamental discoveries about the chemical nature of the atmosphere date to 1687, when Robert Hooke first recognized the significance of the atmosphere as a source of chemical nutrients for plants.)

Remarkably, acid rain as a product of industrial air pollution was discovered and described in the nineteenth century by Robert Angus Smith, general inspector of alkali works for the British government. In 1872, Smith published a book called *Air and Rain: The Beginnings of a Chemical Climatology*. The book addressed such subjects as "Gases of the Atmosphere," "Air of London," "Air of Impure Places," and "Bad Air and the Sensations." Smith identified the phenomenon of acid rain and called it just that: "acid rain."

"Acidity," he wrote, "is caused almost entirely by sulphuric acid, which may come from coal or the oxidation of sulphur compounds from decomposition. . . . It becomes clear from the experiments that rain-water in town districts, even a few miles distant from town, is not pure water for drinking. . . . The presence of free sulphuric acid in the air sufficiently explains the fading of colours in prints and dyed goods, the rusting of metals, and the rotting of blinds. . . . It has been observed that the lower portions of projecting stones in buildings were more apt to crumble than the upper; as the rain falls down and lodges there, and by degrees evaporates, the acid will be left and the action on the stone much increased."

Virtually no attention was paid to Smith's findings, and it wasn't until the mid–twentieth century that another scientist again "discovered" acid rain, quite by accident. In December 1952, when the deadly fog settled over London, one of those present was a young Canadian ecologist named Eville Gorham, who was studying and teaching at the University of London. Gorham was to become this century's pioneer in acid-rain research, although he certainly did not suspect as much at the time.

In the mid-1950s, Gorham moved his family from London to the Lake District of England, just south of Scotland and adjoining the Irish Sea. There, as a research scientist on the staff of the Freshwater Biological Association, he found himself in a happy circumstance: he was able to indulge a longtime fascination with the ecology of peat bogs, an interest that had begun during his boyhood in Nova Scotia.

The unique ecosystems of the world's peatlands are primarily concentrated in the Soviet Union, Scandinavia, Great Britain, Ireland, Canada, and the United States. Until very recently, when large energy corporations began to realize that substantial amounts of energy might be extracted from peat, there was little "practical," commercial reason to study the bogs, so virtually all the extant work to unlock their secrets had been conducted by a handful of scientists like Gorham, who were looking for little more than answers to questions that intrigued them.

Peat, the spongy, decayed matter that underlies the vegetation in bogs, has been described as "geologically young coal." In large bogs, it can reach depths of several feet; if properly treated, it can either be dried and burned efficiently, to produce heat energy, or converted into synthetic natural gas. For an idea of the scope of this emerging source: the energy potential in the reserves of peat in the United States is greater than that of the combined reserves of both oil and natural gas and is exceeded only by that of coal. Energy companies have already hatched plans to mine peat reserves. A few scientists, including Eville Gorham, have expressed grave reservations about the ecological damage that could result from large-scale, strip-mining-style disruption of these sometimes massive wetlands, but in general, the world's bogs are seen as "just swamps," and scientists like Gorham as alarmists or eccentrics.

A bog, however, is not just an ordinary swamp; it is essentially a giant botanical sponge. In 1980, I took a helicopter with Gorham into the heart of the largest contiguous peatland in the continental United States—a deserted, million-acre region in far northern Minnesota known as the Big Bog. From the air, the bog appears to be dry land, pastures of grassy sedges dotted with islands of spindly and scraggy trees that stretch endlessly in all directions. But occasionally, at just the right angle, the sun will glint from a patch of water at the base of the sedges, and when the helicopter has landed and one finally steps out into shin-deep water, it is clear that the bog is a wetland.

For anyone but a field naturalist or wetland ecologist, it might not seem an inviting place, nor the place for a new revelation about the nature of atmospheric chemistry and the effects of air pollution on the rain. But such a revelation did indeed come in a very similar bog in the English Lake District. The key was Gorham's curiosity about the way in which a "true bog" receives its nourishment. He knew that those configurations in the peatlands sometimes called the fens received their nutrients from the mineral-rich water flowing through the spongy peat, in much the same way that plants in coastal saltwater marshes receive nutrients from the sea. But the true bogs were elevated regions in the peatlands, elevated beyond the reach of the mineral-rich groundwaters that flow under and nourish the fens.

Living in and on the true bogs were some strange plants—such as insect-eating pitcher plants and sundews—that received most of their essential nutrients by pulling a switch on Mother Nature: that is, by killing and consuming animals. But this is a rarity. So, what of the most ubiquitous plant in the bog, the mosslike sphagnum? During my visit to the Big Bog, Gorham was continually slogging through the muck and water and pulling up handfuls of the stuff. He had names for the species: *fuscum*, *papillosum*, and *magellanicum*. Sphagnum grows like wildfire in the bogs, and its decay is one of the primary ingredients in peat, so it was abundantly clear to Gorham that these unique plants were receiving nutrition from *somewhere*. But where?

Gorham's only logical response was to turn his eyes skyward. In the few years before his tenure in the Lake District, bog researchers had come to accept that the raised bogs were ombrotrophic, or "rain-nourished." Two scientists in Sweden, Margareta Witting and Einar DuRietz, had been comparing the chemistry of water in bogs with that in various types of fens. Gorham traveled to Sweden to consult with the two scientists.

"I had an interest in the nutrition of sphagnum bogs," Gorham says. "That was all. It was a purely academic interest, of course. I went to Sweden where Margareta was doing water-chemistry studies of bog pools, and Einar was saying that bog pools are essentially reflecting what was coming out of the atmosphere. So when I got back from Sweden, I thought, 'Well, it'd be nice to compare the bog water with what's actually in the rain.' "

Gorham began to collect rainfall in the Lake District peatlands and to analyze the chemical makeup of that precipitation. He had sought only to make a comparison between the chemistry of the rain and that of the bog waters, to do a "major ion balance" of the water coming into the bog from the skies. Thus, his measurements would include a full range of important chemical parameters: sodium, potassium, calcium, magnesium, alkalinity, sulfate, chloride, nitrate, and pH.

To his surprise, he found something he hadn't been looking for. "The minute I started analyzing the rain," he says, "I found that we were alternately dosed with sea salt when the winds blew from the Irish Sea and with acid from the winds that blew up from industrial Lancaster."

Gorham then made a logical leap and postulated that the acid in the rainfall was the result of air-pollutant gases combining with atmospheric moisture, and in 1955, he published what *should* have been a landmark scientific study, entitled "On the Acidity and Salinity of Rain."

Eville Gorham's discovery had enormous implications for commerce, government, science, and the citizens of the world at large. Of course, it was printed in highly technical language in a scientific journal, but at least the scientific community could be counted on to respond.

"The paper," says Gorham, "went over with a dull thud. There was simply *no* interest in the subject. No requests for reprints to speak of."

Dr. Ellis Cowling, another leading acid-rain expert and an associate dean for research at North Carolina State University, characterizes Gorham's discovery thusly: "His pioneering research . . . was met by a thundering silence from both the scientific community and the public at large. One plausible explanation is that Gorham's work, being highly interdisciplinary, was published in a diverse array of scientific journals. In any event, because Gorham's work was not recognized, there resulted a further lag in both the scientific and public awareness of acid precipitation."

Thus, the history of acid-rain research contains two false starts—Smith's and Gorham's. Luckily, there was yet a third thread of discovery, one that began to unravel a decade *before* Gorham began collecting raindrops in the bogs but that did not wind to its end until more than a decade after Gorham published his findings. In the mid-1940s, Hans Egnér, a Swedish soil scientist, conceived a rain monitoring program in Scandinavia to develop a better understanding of the then new concept of rain fertilization of crops. His program was simple: sampling buckets were set up and routinely emptied at experimental farms around the Swedish countryside. Everything that fell into those buckets, dust as well as precipitation, was evaluated according to dozens of routine chemical procedures. Egnér's notion apparently captured the imagination of other agricultural scientists, and by the mid-1950s, the sampling network had spread throughout Scandinavia, then on to most of Western and Central Europe, including the Soviet Union and Poland. (The collection network remains in operation with more than one hundred sampling stations throughout Europe. Although a similar extensive network finally was established in the United States in the 1980s, the European network is the *only* rain-sampling program that has operated continuously and over a long term anywhere in the world.)

The precipitation network established by Egnér allowed two other Swedish scientists, Carl Gustav Rossby and Erik Eriksson, to develop the foundation of the science of atmospheric chemistry. Their notion was that a wide range of chemicals could, like water, be vacuumed into the atmosphere, transported hundreds of miles, and then deposited back on earth. In the 1950s, the two scientists sponsored a series of conferences in Europe to consider these novel ideas (which have by now become established scientific principle). Eville Gorham was one of the scientists attracted to these conferences.

Another was a young Swedish agricultural scientist named Svante Odén. The two young men never met at those conferences; if they had, the acid-rain problem might have been "rediscovered" even sooner than it was. But at the time, Gorham was hardly trumpeting his findings around the European scientific community. Gorham, now an outspoken advocate of acid-rain abatement, compares his own attitude in the 1950s, as well as that of most of his contemporaries, to notions that reached back to the very roots of the industrial revolution: "That was the way it had to be, even with dark clouds from satanic mills hanging over all the cities," he told me in 1982. "When I was doing that early acid-rain research, I had no environmental consciousness. As far as I was concerned, it was just an interesting situation. And it was my job to put it into the scientific journals. So I pursued it as an intellectural problem, not as a pollution problem. I was simply interested in the question, What determines the chemistry of the rain?"

Gorham's work in Great Britain may have been largely an intellectual exercise, but by the time he returned to his native Canada in 1959, he had learned enough to have laid the foundation of present-day understanding of the phenomenon, including a study published in 1959 that showed a clear correlation between the incidence of bronchitis in humans and the acidity of precipitation.

The path followed by Svante Odén took a different course. In 1961, he began collecting samples from Scandinavian lakes and rivers as part of an effort to establish a surface-water-monitoring network, but as his data accumulated, he began to see a regular series of trends in lake and river chemistry that related directly to data being gathered by the *rain*-collection network. Odén suddenly found himself evaluating the history of air-pollution emissions in various regions of Europe, the amount of time any pollutant could remain suspended in the atmosphere, the distance and direction it might travel, and the sorts of chemical changes it might undergo if suspended long enough.

The evidence he assembled indicated that when major weather systems moved over Scandinavia from such heavily industrialized regions of Europe as the steel-producing Ruhr Valley in Germany and the industrial heart of Great Britain, acid precipitation fell on Sweden and Norway.

Further, Odén demonstrated that the acidity of the rain over northern Europe had increased during the years the European network was in operation, and he showed a clear trend between the increase in acidity and the increase in emissions from fossil-fuel combustion in the industrial heart of Central Europe. He also pointed out that as the acidity of rain over Scandinavia had increased, so had the acidity of Scandinavian surface waters.

Odén went on to suggest that if his conclusions about acid rain were correct, this troubling new ecological problem could lead to a wide range of problems: acid might leach molecules of toxic metals out of soils; acid could leach essential plant nutrients away; acids could destroy fish populations, impoverish the soils of forests, increase the incidence of disease in plants, and even damage man-made structures of marble and limestone.

Rather than discreetly submitting a paper to a relatively obscure scientific journal, Odén published his findings on October 24, 1967, in the Swedish newspaper *Dagens Nyheter*.

Today, Eville Gorham says that it was Odén's bold move that at last catalyzed action in both the scientific and political communities. "It seems to me that part of the problem was that we [as scientists] never went out on the stump."

Largely as a result of Odén's work, the Swedish government in 1972 released the results of its own inquiry into acid precipitation. The report was presented in Stockholm, at the United Nations Conference on the Human Environment, a landmark event and an outgrowth of a burgeoning worldwide awareness of environmental issues. The report agreed in large measure with Odén's findings: massive fish-kills had occurred in hundreds of Scandinavian lakes and streams; acid precipitation was quite probably a threat to agricultural crops and the commercially valuable forests of northern Europe, to monuments and other property, to human health. And, said the Swedish report, which was printed in English, "A similar situation might possibly exist within certain regions of Canada and the northeastern part of the U.S.A. A detailed study of the likelihood of such a development is a matter of urgency."

If such a study was "a matter of urgency" for North America, did the U.S. Congress and the Canadian Parliament swing into action to order such research? Did the then youthful U.S. Environmental Protection Agency or its Canadian counterpart initiate such a study on its own? Or did the American news media, always hungry for a new angle, leap to investigate this "matter of urgency"? They did not.

But not everyone ignored the message. During 1971, Odén toured the United States, presenting a series of fourteen lectures about his findings, and in March of 1972, American biologist Gene E. Likens of Cornell University, along with F. Herbert Bormann of Yale and Noye M. Johnson of Dartmouth, published a report in the journal *Environment*. Entitled simply "Acid Rain," the article pointed to the scarcity of data on rain chemistry over the years in the United States. "We can find no data to accurately describe long-term trends for the U.S."

But the three men had become aware of an apparent problem with the rain while doing studies of "biogeochemical" cycles—that is, attempting to measure all the biological and chemical influences entering and affecting the Hubbard Brook Experimental Forest in New Hampshire. Their precipitation data showed annual average pH readings of between 4.03 and 4.19 in the years between 1965 and 1971, and an all-time low reading of a "surprisingly acid" pH 3.0. The scientists noted that low pH values were being recorded elsewhere in the northeastern United States: annual pH values around 4.3 in New Durham, New Hampshire; Hubbardston, Massachusetts; and Thomaston, Connecticut; a summer 1970 average of pH 3.62 in New Haven, Connecticut.

Although Likens and his colleagues lamented the lack of historical data, they did unearth some records of rain samples from Geneva, New York, obtained during the 1920s, and although there were no pH data from those samples, there were measurements indicating large amounts of bicarbonate. "Bicarbonate," they noted, "cannot exist with the stronger acids found in today's rain. The presence of bicarbonate, therefore, would indicate that pH values in 1919–1929 were 5.7 or higher." Further, the scientists found that glacial ice samples taken in the northern Cascade Mountains of the northwestern United States showed historical pH levels of about 5.6.

For Gene Likens, the article was the first of a series of four published in the 1970s, culminating with a report in *Scientific American* in 1979 (with coauthors Richard F. Wright, James

N. Galloway, and Thomas J. Butler). The latter article summed up the findings of Likens and various colleagues: that acid rain was not only falling over portions of North America but steadily expanding its influence over wider and wider regions; that in Europe, acidity of precipitation at some monitoring stations had increased tenfold since 1955, exactly paralleling increases in sulfur-dioxide and nitrogen-oxide emissions; that even though historical data in North America were not as complete as in Europe, there appeared to be "intensification of acidity" in the eastern United States (in addition to the southward and westward extension of the regions receiving acid rain); that available evidence gleaned from ice cores taken from Greenland's glaciers suggested that two hundred years ago, at the dawn of the Industrial Revolution, rainwater was close to neutral.

But perhaps the most alarming piece of research came from a colleague of Likens's at Cornell. Likens had pointed to the skies, suggesting that the rains had become so acidic that some North American ecosystems might be threatened, and in 1975, biologist Carl Schofield examined the status of 214 high-altitude lakes in the Adirondack Mountains. Throughout the 1960s and 1970s, the New York Department of Environmental Conservation had become increasingly concerned about the disappearance of trout from a smattering of those high-elevation wilderness lakes. Yearly, the agency's fisheries biologists had dumped thousands of healthy trout fingerlings into lakes where trout had once thrived, but none were surviving.

Carl Schofield's report stated that 82 of the 214 lakes were no longer capable of supporting fish life, for they had become too acid, and more than two dozen of the remainder were on the verge of becoming too acidic for effective fish reproduction. The cause appeared to be polluted rain.

7

The little town of Cheshire, Ohio, sits among cornfields and gorgeous wooded rolling hills of the Ohio River Valley. On the riverbank just west of town, wild white morning glories trail along the gravelly shore, and on the brown, turbid waters of the Ohio River, a barge, piled high with nuggets of coal, nudges its way upstream. The boat pushing the barge is white and blocky, and if one stretches the imagination just a bit, it has the hint of Samuel Clemens and riverboat about it.

From the riverbank, even with the whining and rumbling of automobiles and trucks nearby on Highway 7, the scene is pastoral. But here, just outside peaceful little Cheshire, lies the epicenter of the acid-rain problem.

The nub of the acid-rain story is neither atmospheric chemistry nor freshwater biology, neither geology nor meteorology. Acid rain continues to be a problem even though equipment to control it has already been developed, and the reason for this, purely and simply, is politics. Politics allows acid rain to continue. But even worse, and more bizarre, politics not only condones acid rain but sometimes even encourages industry to exacerbate the problem. Here in Cheshire, the political tangle surrounding acid rain becomes more vivid than anywhere else.

As you stand on the riverbank, turn upstream, and you'll see, just north, a sky lanced by two towering industrial smokestacks. The stack in the foreground belongs to an electric-power plant called Kyger Creek, a coal burner operated by the Ohio Valley Electric Corporation. The taller stack and the two, huge, hyperbolic cooling towers in the background belong to another, even more massive coal burner, the General James M. Gavin plant, operated by the Ohio Electric Company, which is a subsidiary of the Ohio Power Company, which is a subsidiary of the American Electric Power Corporation, the largest for-profit electrical company in the world.

These are but two of hundreds of similar coal burners dotted across North America, two among dozens that stretch along the Ohio River as it winds through Ohio, West Virginia, Indiana, Kentucky, and Illinois. Virtually all these coal-fired plants belch some measure of sulfur dioxide and nitrogen oxides into the atmosphere over the United States, but of the twenty greatest sulfur polluters in the U.S., *seventeen* lie in this same geographic neighborhood, in a three-hundred-mile radius around the city of Dayton, Ohio.

In 1980, Kyger Creek was the fifteenth-greatest emitter of sulfur dioxide in the United States; Gavin was number one. And since the two plants stand right beside each other, their combined emissions make this spot outside Cheshire the single worst square mile for sulfur-dioxide discharge, *and* one of the worst for nitrogen oxides, in the United States. More than 570,000 tons of sulfur dioxide alone spew from the two tall smokestacks annually. (This astonishing combined discharge, however, still pales beside the emission from the world's greatest sulfur-dioxide source, the Inco nickel smelter in Sudbury, Ontario.)

Those 570,000 tons, along with the millions of yearly tons from other huge coal plants and other sulfur-dioxide sources, are clearly the primary cause of acid rain. The fact that these emissions must be controlled if acid rain is to be checked is recognized by almost every acid-rain researcher in the world, at least by those not in the direct employ of an electric-utility company. And until the Ronald Reagan/James Watt nexus took over the United States' natural-resources and environmental policy, it was a fact that had been recognized by most environmental overseers at the national level.

So it is a remarkable irony that long before the arrival of Reagan/Watt, Gavin and Kyger Creek and hundreds of other plants just like them were polluting the air with the full knowledge and blessing of the United States Congress and of the U.S. and Ohio EPAs. Despite massive emissions of the chemicals directly responsible for acid rain, neither Kyger Creek nor Gavin is in violation of any federal or state environmental law. Further, both plants are equipped with devices that appear to make acid rain much worse. And these devices—the tall smokestacks attached to both plants—were installed at the encouragement of the EPA.

From the roof of the Gavin plant, some three hundred feet up, the guide from the Ohio Electric Company and I looked out over the river valley. Directly below us was Gavin's food supply: a two-million-ton pile of coal, enough to feed Gavin for a hundred days. The pair of hyperbolic water-cooling towers straddled either side of the plant, each large enough at the base to contain a major-league baseball diamond. Thick columns of cloudy white steam billowed from the towers, and above-them, a thinner, just-off-white trail of emission gases poured from the tip of Gavin's thousand-plus-foot smokestack. We were actually standing on one of two twenty-five-story buildings that house Gavin's twin boilers and that hug tight beside the tall stack like testicles.

Beneath our feet, Gavin's roof shuddered and trembled. We stepped back inside and stood on a steel-mesh floor, looking down through another twenty-plus steel-mesh floors to the base of the boiler far below. On each level, dim yellow light bulbs seemed to swim in midair.

What happens inside Gavin is similar to what happens at almost any other huge coal plant anywhere in the world. Coal, the distant fossil remains of life from millions of years ago, pours into the plant by conveyor at the rate of 860 tons per hour. In Gavin's case, about one third of the coal comes from a "captive" coal mine a few miles away. The rest of the plant's 7.5-million-ton annual diet arrives by river barge at the plant's own docking and unloading facilities. Once inside the plant, the coal feeds into a hardened-steel pulverizer, to be hammered and ground to the consistency of baby powder, then is mixed with air from a series of powerful fans and blown into the fiery maw of the twenty-five-story boilers. The atomized coal explodes into flame. Purified water flowing through tubes in the boiler transforms into superheated steam: 10 million pounds of steam per hour to blast against the blades of mechanical turbines at 3,500 pounds per square inch pressure thus spinning the blades at supersonic speeds, to rotate huge magnets inside massive coils of heavy copper wire in the generators; billions upon billions of electrons are driven off; and electric current made indirectly from the tissues of prehistoric life flows at a rate of 2.6 billion watts out of Gavin and on to Ohio Power's 600,000 customers.

Citizens of the United States are the world's greatest per capita consumers of electric power, and of energy in general. In 1978, the United Nations Statistical Yearbook noted that each American consumed the energy equivalent of close to 12,000 kilograms of coal per year. Energy consumption in such high-living-standard European nations as Germany, Denmark, and Switzerland was roughly half that level.

Between 1965 and 1978, the U.S. demand for electricity more than doubled, from 3.3 quadrillion Btu's (British thermal units) per year to 6.8 quadrillion Btu's. According to the U.S. Department of Energy, that figure is expected to nearly double *again* by 1995, prodded along by a growing population, decreasing availability of oil, and sharply rising prices for natural gas. Most of that electricity will come from coal burned in plants like Gavin. In fact, by 1995, the Energy Department predicts that about 42 percent of *all* U.S. energy consumption will come from coal burned to produce electricity. (The greatest share of the remainder will be petroleum used to power automobiles and trucks.)

Thus, in America's future coal will be king, and in the service of king coal, an estimated 350 new coal-fired power plants, virtually all of them big ones, will be built before 1995. Unless the lobbyists hired by utilities and other polluting industries are extremely successful at gutting the present U.S. Clean Air Act, every one of those new power plants will be equipped with devices that help control sulfur dioxide—these devices, "scrubbers," spray a neutralizing solution of dissolved lime or similar substance into the emission stream before it enters the smokestack. Much of the sulfur dioxide will bond chemically with the dissolved lime to form a thick, mucky slurry of calcium sulfate, which can be removed from the scrubber. These scrubbers can remove 90 percent or more of the sulfur dioxide from any plant's emissions. The best of the scrubbers will remove more than 95 percent, and, perhaps, as technology improves in coming years, even closer to 100 percent.

Regulations developed by the EPA and required by the Clean Air Act amendments adopted in 1977 will see to it that such devices are installed and operated in all 350 of these new plants. But, remarkably, the EPA predicts that sulfur-dioxide levels will continue to rise, at least until 1995, and Gavin and Kyger Creek play a key role in the reasons for this increase.

When I visited Gavin, I had come primarily to examine what the plant and its operators do *not* do to protect the atmosphere, but my guide was eager to show me what Gavin *does* do. When we stood on the roof, he pointed out a series of air-pollution-control devices—boxy, steel-sided structures—between the boilers and the huge stack. Standing atop such a massive plant, one can easily lose perspective, but each of the dozen devices is the size of a small summer cottage. They are called "electrostatic precipitators," and they are in wide use almost everywhere to control some of the most obvious industrial emissions. Certainly, of all air pollutants, the most noticeable are the trillions of motes of dust, soot, and ash by-products of combustion that would normally spew from an industrial stack as clouds of thick black, brown, or orange smoke. Electrostatic precipitators remove these particulates with remarkable efficiency. In an electrostatic precipitator, a high-energy electrical field charges the particulates so that they can be filtered from the emission stream by electromagnetic attraction. A series of precipitators is often necessary, since each device will remove only a certain fraction of the particulates in the emission: the first device might remove 60 percent of the particulates, the next 20 percent, the next 6 percent, and so on. There is no dispute that Gavin's operators do a fine job of particulate control: well over 99 percent of Gavin's dust and soot is removed from the emission stream. Thus, the emission pouring from the stack is nearly pure white.

Further, the two cooling towers help protect the thermal quality of the Ohio River, beside which the plant is located both for the abundant supply of water and for the river's convenient role as a highway for coal shipping. The towers allow the plant to operate a nearly closed-cycle water system. Steam from the boilers is condensed in the towers and recycled back into the system cooled and purified, instead of being dumped hot back into the river.

Ohio Power's own brochure on the Gavin plant states, "Total cost of the plant was more than $600 million. Approximately 15 percent of this cost was spent for environmental protection facilities. While this equipment does not produce any revenue and actually reduces the capacity and efficiency of the plant, the expenditure was necessary to maintain environmental quality." The brochure does not point out that most of the environmental expenditures were incurred over the objection of Ohio Power, because of requirements of the U.S. Environmental Protection Agency. Nor does it mention that the plant has virtually no equipment to remove sulfur dioxide from its emission stream, despite the fact that such pollution-control equipment already existed at the time Gavin was built. Less than a decade old at this writing, Gavin almost surely will be operating another thirty years or more—far into the twenty-first century—so for decades to come, coal laden with sulfur will continue to be atomized to fine powder and fanned into Gavin's boilers. In the flames, the elemental sulfur will be oxidized to sulfur dioxide, to pour out of the plant's chimney at rates measured in tons per hour. To meet certain relatively lax state and federal standards, Gavin's operators do, at least, burn a mix of high-sulfur and more expensive low-sulfur coal, and the company has begun to convert some high-sulfur coal to lower sulfur content by "washing" it. But as the law now stands, the company need do nothing to remove even a trace of sulfur dioxide from the plant's emission stream.

Yet according to Ohio Power, the company does indeed "control" sulfur dioxide, and at a substantial "unproductive" cost to the company. If one were to ask a guide, during a tour of the plant, precisely how this sulfur-dioxide "control" is accomplished, chances are he would simply point skyward. "The single 1103-foot concrete and steel-lined stack puts gaseous emissions into the upper atmosphere where they are diluted, dispersed and disseminated so as not to form harmful concentrations at ground level," says the Gavin handout.

The entire acid-rain story is fraught with irony, and this may be the most pointed of them all: Gavin's tall stack was built to help prevent the buildup of high, ground-level sulfur-dioxide concentrations near the plant in order to protect public health. Yet by injecting its gaseous offal into the "upper atmosphere," the big stack becomes a major factor in the acid-rain problem, for it sends pollutants high enough to be carried extraordinary distances, and altitude provides these gases with the time they need to interact with other chemicals to form ecologically deadly acids.

Understanding the federal and state policies that allow plants like Gavin to pollute with relative impunity requires a trek into the regulatory swamps of government, especially the mire surrounding the U.S. Clean Air Act. The Act is a tortuous bit of law that has produced carloads of related regulatory schemes, particularly the so-called State Implementation Plans. As part of the Act, each of the individual states was required to prepare such implementation plans for each of several "criteria pollutants." The idea was that each state would prepare its plan under the close supervision of EPA, so that federal pollutant-level standards could be met in a manner most efficient for each state. But as a result of the multilevel nature of the program, the swamp became almost impenetrable. The State Implementation Plans (SIPs) and related programs became largely a mire of acronyms that only an unrepentant bureaucrat could truly appreciate.

One of the most delightful bits of unintentional comedy came from EPA requirements that the state pollution regulators *explain* their Act-related programs to the citizenry, as part of an all-out effort for "public participation." I attended such a meeting in 1977, shortly after that year's new amendments to the Act were passed, and found a roomful of local elected officials and industry air-pollution specialists listening to a Minnesota official explaining the ramifications of the Act. He spoke of the SIP to be completed to meet the NAAQS; the PSD requirements and CO problems to be resolved, especially in the CBDs. New plants would have to install BPCT or BACT in order to meet the NSPS. At one point, a local elected official attempted to wring a clarification out of the state spokesman, and was met by a further, but very calmly delivered, blizzard of acronyms and double-speak. The local official glanced around the room, a look of fleeting panic on his face, then nodded at the state spokesman and, apparently punch-drunk, settled back in his chair to wait out the end of the seminar.

The swamp notwithstanding, the basics of the acid-rain problem and its relationship to the Clean Air Act can be understood by looking at two key elements. Let's call them (1) the westerly wind joke and (2) sources old, sources new.

There was a running gag at the Minnesota Pollution Control Agency that in the minds of many of those responsible for issuing industrial air-pollution permits, the best air-pollution device ever invented was "a strong westerly wind."

The joke isn't considered funny in Wisconsin, which lies about seventeen miles east of downtown St. Paul. But the joke has roots sunk deep in reality, for the strong wind—westerly or not—has been at the core of the traditional engineering concept of how to control air pollution. (For that matter, a similar principle has always been applied to pollution control in general. In the case of water pollution from, say, sewage, the law of the land has routinely allowed enormous amounts of waste to be poured into a waterway as long as the waterway's flow is sufficient to dilute the effluvium to an "acceptable" level.)

When Congress assembled the Clean Air Act in 1970, it had two objectives: to protect public health and to protect public "welfare," the latter meaning property, recreational resources, crops, wildlife, and other "nonhealth" aspects. Congress ordered the newborn Environmental Protection Agency to establish "primary" standards to protect human health and, if necessary, even tighter "secondary" standards to protect property and natural resources. These would be the National Ambient Air Quality Standards.

There's the rub. The EPA proceeded to develop a series of "criteria documents" for a handful of important air pollutants—sulfur dioxide, nitrogen dioxide, ozone, carbon monoxide, and particulates. These documents were to provide the EPA with scientific evidence to set appropriate primary and secondary limits for the concentrations of these pollutants in the air, but *not* the concentrations of the pollutants as they roared out of smokestacks. Rather, *ambient* standards were to be based on the concentration of pollutants in a sample of air taken at human-nose level. For example, if the EPA determined that more than one part per million of sulfur dioxide in the air was dangerous to human health, it could establish its ambient standard at a level below one part per million. Then, to determine whether that standard was exceeded, EPA and its state counterparts would attempt to measure samples of ground-level air at various locations—particularly those locations nearest industrial and urban centers. If the air sample exceeded the health-related standard (welfare-related standards have almost never been enforced), then the pollution-regulating agencies would attempt to track down the offending polluters through the use of sophisticated computer air-dispersion models. If the offending polluter or polluters were found, and if the regulators could somehow prove that those polluters were responsible for the exceeded standards, *then* the polluters could be required to find a way to meet the ambient standard.

Enter the westerly wind. In many cases, all the polluter had to do to resolve the ambient-air-quality violation was to find a way to blow the pollution away from the regulator's monitors, or to dilute it to such a degree that no violation was recorded. In other words, hundreds of polluters, with the blessings of the very federal and state agencies responsible for controlling air pollution, were allowed to *continue* pouring as much crud into the atmosphere as they could. All they had to do—and precisely what the regulators openly advised them to do—was to build a tall smokestack. There was nothing nefarious about this, for it was accepted engineering practice: if the emissions were massive, the answer was to build an extraordinarily massive chimney.

Without question, big chimneys *are* effective diluters of pollution, but they do not actually abate the tons of air pollution that might spew from a plant like Gavin. And since the atmosphere is a closed, continually self-cleansing system, there was never any real question that the diluted emissions would eventually return to earth.

In fact, it now appears that tall stacks do a great deal to exacerbate the acid-rain problem. Because there are abundant supplies of natural alkaline buffers in the watersheds of the Ohio River Valley, there is virtually no chance that emissions from a plant like Gavin or Kyger Creek could harm a *local* lake or river. (''There isn't enough coal in the world to acidify those waters,'' an EPA scientist told me.) But the huge chimneys help gaseous pollutants to travel far and wide, and into regions that *are* acid-sensitive. Further, they give those pollutants plenty of "residence time" in the atmosphere—the days and hours they need for the slow atmospheric brew into acids. (About 1 percent of the sulfur dioxide in a given parcel of air will transform into acids each hour. Obviously, the longer the gas remains suspended, the more opportunity it has to convert to acid.)

Increasing chimney height has been an accepted pollution-control technique since the early days of the Industrial Revolution. After its Killer Fog of 1954, London dealt with much of its air-pollution problem by requiring industries to install tall stacks. Now the chances of such an episode are remote, but now, too, Great Britain is the second-greatest source of acids that come to rest in Scandinavia.

In 1955, there were only two stacks in the United States taller than 600 feet; by 1980, there were 175 stacks of over 600 feet, and at least fifteen stacks taller than 1,000 feet worldwide. The tallest in the world, at the Inco Ltd. nickel smelter in Sudbury, is nearly a quarter-mile high. Its closest U.S. competitors include a 1,206-foot stack at an American Electric Power Company plant in West Virginia and Kennecott Copper's 1,200-foot Magna Smelter chimney in Utah.

Electric-utility companies have been promoters of tall stacks for years. Gus Speth, who was chairman of the Council on Environmental Quality under President Jimmy Carter, told an audience at a 1979 Canadian conference on acid rain, "One electric utility [American Electric Power, now the operator of Gavin] went so far as to take out newspaper and magazine ads back in 1973 bragging that it was a 'pioneer' in the use of tall smokestacks on its power plants to 'disperse gaseous emissions widely in the atmosphere so that ground level concentrations would not be harmful to human health or property.' The company claimed that gases from tall smokestacks 'are dissipated high in the atmosphere, dispersed over a wide area, and come down finally in harmless traces.' It went on to blast what the company called 'irresponsible environmentalists' who insisted on tough emission standards . . . charging that they were guilty of 'taking food from the mouths of people to give them a better view of the mountain.' "

The EPA, as well as many of its state counterparts, has always been willing to accept tall stacks as part of a "cost effective" approach to air-pollution control. In 1974, the Natural Resources Defense Council successfully sued EPA for allowing Georgia to base much of the state's pollution-control program on the widespread use of tall stacks. As recently as 1980, the Minnesota Pollution Control Agency approved a plan by a refinery just south of Minneapolis to control its sulfur dioxide by building a taller stack—in the same year that the same agency was voicing alarm at studies indicating that the state's northern lakes were threatened by acid rain!

One of the great achievements of the Clean Air Act of 1970 was that Congress, over the vehement objections of the polluting industries, at least attempted to turn the United States away from regulations based on the westerly wind joke and in the direction of *emission-based* air-pollution standards: that is, to put a limit on the actual amounts of pollutants that could spew from any industrial chimney, short or tall.

Enter the second trail through the political swamp: sources old, sources new.

In 1970, Congress was faced with a political dilemma. The pressure was on from environmentalists to turn away from nose-level standards and instead compel industries to install the best emission-control equipment available. Industry lobbyists, on the other hand, insisted that any requirement to outfit all polluting plants and factories with such equipment would be economically crippling. So Congress struck a compromise. Industries were ordered to meet federal emission limits, but all "old," *existing* industrial facilities were to be completely exempt from those limits. Only brand-new plants and factories would be governed by the tight emission controls.

In the case of coal-burning power plants, this meant that any new facility would have to meet a fairly rigorous standard, limiting its emissions of sulfur to 1.2 pounds for every 1 million Btu's of energy generated. Typically, it was a tortuously complex scheme, since sulfur content of any batch of coal will vary widely from the next. In a very general sense, the effect was to cut emissions by roughly 50 percent. To meet these limits, operators of new plants would almost certainly have to install sulfur-dioxide scrubbers. But existing, or "old," power plants merely had to comply with the old, nose-level, ambient standards. And as far as the Clean Air Act was concerned, an "old" power plant was any under construction by 1971. (Technically, the Clean Air Act also required these old plants to meet emissions standards chosen by each state. These tend to vary widely. In 1980, they ranged from Connecticut's .55 pounds per million Btu's to Ohio's 9.9 pounds per million.)

In its historical framework, the significance of this policy is enormous. During approximately the same period that stack heights were increasing (between 1950 and 1980), there was also a sharp shift in the sources of sulfur dioxide. Before World War II, most sulfur dioxide came from residential and small commercial boilers with only rooftop chimneys. In 1940, Americans emitted roughly 20 million tons of sulfur dioxide annually, but only 3 million of that total came from power plants. Today, approximately 30 million tons of sulfur dioxide are emitted yearly. And nearly *two thirds* of that total comes from coal-burning electrical-power plants, and the vast majority of that two thirds comes from plants that are classified as "old" sources not subject to the emissions standards of the Act.

Consider the Gavin plant. Even though it was not completed until 1975, it was classified as an old plant. If scrubbers had been installed at Gavin in 1975, its output of airborne pollutants would be less than half the present level.

By 1980, standards for new coal burners had been tightened and simplified, calling for new power plants to remove at least 90 percent of sulphur when high-sulfur coal is burned and 70 percent when lower sulfur coal is burned. In effect, the standards require all new coal burners to be equipped with scrubbers. (However, there is nothing in the regulations to prevent a progressive utility from using an even newer, less expensive technology than scrubbers should it be able to develop one.)

The gap between requirements for old and new sources means that a plant like Gavin can legally pump out eight times as much sulfur dioxide as could a new facility of the same size. Ohio alone has twenty-one such major old plants and, as a result, emits an average of 64 tons of sulfur dioxide for every square mile of the state annually. (Again, Ohio is not a solitary culprit. West Virginia emits 43.3 tons per square mile; Pennsylvania, 33.1.)

Without question, the New Source Performance Standards eventually will yield an improvement in the amount of sulfur discharged into the environment, but not in the near future—not, in fact, until near or even after the turn of the century.

The EPA estimates that sulfur-dioxide emissions from utilities will increase to at least 20.5 million tons by 1995, roughly a 10 percent increase from the 18.6-million-ton level of 1975 (the year Gavin was completed). Of these 20.5 million tons, as much as 75 percent will be coming from the old power plants.

Says attorney Robert Rauch, formerly the leading legal expert on acid rain for the Environmental Defense Fund, "This is the anomaly that is at the heart of the acid-rain problem. Until these sources are controlled, we're not going to see *any* appreciable change in the acid-rain problem. Despite all the rhetoric, the reality is that the situation is not even stable. The situation is deteriorating."

The intention of Congress when it distinguished between "old" and "new" sources was to avoid burdening operators of existing power plants and industrial facilities with expensive pollution-control requirements. In some cases—particularly in the oldest of the old plants—requiring scrubbers at costs in the tens of million might not make sense if the plant is to be retired in the next few years. But the typical life span of a coal burner is forty years or more, and installing scrubbers at a plant like Gavin would make plenty of sense. There is no question that Gavin is much cheaper to operate without sulfur-dioxide-control equipment, but there is also no question that Gavin will continue to spew its emissions for every moment of its forty-year, or longer, life. Further, under current law, utilities are free to shut off their cleanest plants when electrical demand is low and instead operate their older, dirtier, and cheaper facilities at full tilt. (This is not always done. Wayne Kaplan, of Minnesota's Northern States Power Company, says that NSP routinely operates its huge SHERCO coal-fired facility as much as possible. The plant is equipped with scrubbers, and in this case because of greater design and operating efficiency, the plant is actually cheaper to run than the company's older and much dirtier coal burners.)

Most researchers agree that sulfur dioxide is responsible for about two thirds of the acidity of rain in the eastern United States and Canada, the oxides of nitrogen for the remaining third. In the western United States, the mix appears to be closer to half and half, largely because much less sulfur dioxide is emitted in the west.

The control of nitrogen oxides presents a different set of

problems. About 30 percent comes, once again, from the electrical utilities, but an even greater amount—a full 50 percent—comes from motor vehicles, most prominently the family automobile. The remainder comes from refineries, factories, and even residential and commercial furnaces. In 1978, some 24 million tons of nitrogen oxides poured into the atmosphere of the United States. Most of this is formed when extraordinarily high heat in engines and modern industrial furnaces literally burns natural nitrogen in the air.

Current EPA standards require power plants to reduce uncontrolled nitrogen-oxide emissions by only 20 percent. But it now appears that by 1990, emissions of nitrogen oxides could equal or even exceed sulfur-dioxide emissions, partly as a result of steadily increasing motor-vehicle traffic and partly as a result of a design characteristic of modern power-plant burners. Most new power plants tend to operate at considerably higher temperatures than their predecessors, thus producing more nitrogen oxides. (An incidental but equally disturbing note is that the oxides of nitrogen tend to react with sunlight and pollutant hydrocarbons in the atmosphere to form ozone and other photochemical oxidants—primary ingredients associated with urban smog. Beyond a clear threat to health—virtually all of the frequent air-pollution health alerts in Los Angeles are called because of high ozone levels—recent studies by Dr. Sagar Krupa at the University of Minnesota have indicated that ozone may be destroying up to 15 percent of Minnesota's multimillion-dollar soybean crop. Like the components of acid rain, ozone and other photochemical oxidants travel enormous distances in polluted regional weather systems.)

To date, the thrust of nitrogen-oxide abatement in the United States has been to control emissions from automobiles. In the 1970 Clean Air Act, the goal was to reduce emissions from automobiles to 10 percent of uncontrolled levels, but in 1977, under intense lobbying pressure from the automobile industry, those standards were substantially relaxed. Originally, nitrogen-oxide emissions from cars were not to exceed four tenths of a gram "per vehicle mile," but that tough limit was eased to two grams in 1977, with a requirement that auto makers improve emissions to one gram by 1981. At this writing, the automobile industry is attempting to engineer another relaxation from Congress—back to the pre-1981 levels of two grams per mile.

Meanwhile, as incredible as it may seem, there is nothing in United States law that states that acidifying a lake, stream, or forest community is illegal. Thus, although the Clean Air Act at least attempts to deal with the problem of sulfur-dioxide and nitrogen-oxide emissions, it says nothing about the deposition of sulfates, nitrates, or sulfuric and nitric acids. By 1983, only one of the fifty states—Minnesota—was actually in the process of developing such a standard.

In 1982 and 1983, as Congress began to review the Clean Air Act, which was up for renewal in the depth of a severe economic recession, pressure was heavy to relax even the existing, inadequate standards. A National Coal Association blueprint for improving the Clean Air Act included among its proposals a notion that cost-benefit analysis should be applied in the process of setting ambient health-related standards and that "economic impact statements" be required for new regulations. On the face of it, these ideas might have some merit, but a deeper examination shows that *only* the costs of pollution control can be adequately calculated. It is simple to determine that a sulfur scrubber costs $90 million. But even if some sort of dollar figure could be attached to say, the value of a longer, healthier human life, the possibility of *proving* an absolute link between any specific piece of pollution-control equipment and a specific case of disease or death would be functionally impossible. One need only look at the tobacco industry's continuing attempts to deny links between smoking and cancer to get a sense of the obfuscation and delay that would occur if the industrial polluters could install such a cost-benefit program. Although no one, including the National Coal Association, has yet been able to determine the monetary value of a human life or of a northwoods ecosystem, the approach such lobbies might take toward development of any cost-benefit regulations is probably reflected in the comment of Coal Association president Carl Bagge about existing law: "Because of clean-air regulations, the costs of using coal are much higher than necessary, with few, if any, compensating benefits."

But even with adverse pressure from both industry lobbyists and the Reagan administration, one major effort by Congress to clamp down on the sources of acid rain got under way in 1982. In July, the Senate Environment and Public Works Committee passed a bill sponsored by Sen. George Mitchell of Maine that took direct aim at electric utilities in the eastern U.S. As approved, the Mitchell bill, designed with the aid and support of the National Clean Air Coalition, a consortium of leading environmental, labor, and citizen-action organizations, aimed to reduce sulfur dioxide in the eastern United States by 8 million tons, or about 35 percent, by 1990. Under the provisions of the bill, those reductions were to be required in thirty-one eastern states in a complicated scheme that would mandate the greatest improvement in states with the most lax emission controls.

By 1983, the prospect seemed dim that the bill might pass both houses, and even if it did squeak through Congress, a presidential veto seemed all but certain. Lobbying pressure from the coal and utility industries was intense. To be sure, environmental organizations were also mounting a furious lobbying campaign, but the industry lobby was pointing to the cost of retrofitting hundreds of dirty coal furnaces with scrubbers—a cost estimated by the Congressional Office of Technology Assessment at about $2.5 billion per year, which certainly would be passed along to utility consumers in the thirty-one states in the eastern U.S.

For consumers of electric power in the eastern United States, a 10-million-ton reduction in sulfur emissions would increase utility bills by an average of only 1.9 percent in 1990, according to figures provided to Congress by the National Clean Air Coalition. However, the rate increase would be higher in some areas than in others: 6.5 to 8 percent in high-sulfur-emission states like Ohio and Indiana.

Says the coalition's David Hawkins, a former top EPA air-quality official, "The states with the largest increases are among today's largest emitters. In 1990, even after these rate increases, these states would still enjoy rates less than those in the mid-Atlantic and New England states." (Hawkins adds that the coalition-backed Mitchell bill focuses on the utilities because the utilities emit more than two thirds of the region's sulfur dioxide; because, as the Clean Air Act currently stands, there will be no net reduction in overall utility emissions in the twentieth century; and because the huge utility burners represent a cost-effective target for reductions, since so many are largely uncontrolled.)

Politicians from the eastern coal states were under another sort of pressure. In these states, where unemployment had soared into double digits, any prospect of further restrictions that could affect coal mining was greeted with alarm.

Between World War II and 1970, coal employment in the United States declined some 46 percent, but during the 1970s, there was supposed to be a boom resulting from the deepening crude-oil crisis. That boom came, but most of it occurred in the western United States, where production increased by about 450 percent, whereas eastern production increased by only 10 percent. At least part of the reason for this imbalance was the abundance of low-sulfur coal in the West.

Coal industry spokesmen have testified before Congress that more than 98,000 jobs could be lost in mining in the Appalachian and Midwestern coal regions if the Mitchell bill were to become law. In 1981, the Region 6 local of the United Mine Workers, which covers Ohio and portions of West Virginia, said that 5,000 to 5,500 of the union's members were not working because of layoffs. James Rhodes, then governor of Ohio, was fond of citing this figure as an indication of the number of jobs snatched away by the Clean Air Act, but figures from Rhodes' own Ohio Bureau of Employment told a different story: 15,302 miners were working in 1980 in Ohio, close to double the 1965 figure and only 1,819 fewer than in 1979. In 1979, there were more Ohio coal miners employed than in any year since 1950.

Also, the Coal Association's figure of 98,000 lost jobs is highly questionable. A document issued by the Canadian Embassy in the U.S. points out that this argument assumes that utilities will in all cases immediately switch to the lowest-sulfur coal available. But even if the program resulted in a 67-million-ton coal reduction in Appalachia's high-sulfur-coal areas, another 34 million tons would likely be mined in the region's low-sulfur-coal areas.

However, the Canadians do admit that even if the bill boosted low-sulfur-coal mining in Appalachia and the western United States, and even if manufacture of pollution-control equipment generated thousands of *new* jobs (as it certainly would), there could still be significant job disruption and dislocation in the coal regions.

Still, says the Canadian report, "There should be taken into account the job loss or dislocation that could result if acid rain is not controlled. Thousands of jobs in Canada, New England, New York and Colorado are dependent upon the tourist industry. To the extent that tourism suffers if lakes and rivers become fishless or forest damage occurs, job losses and dislocation could be considerable."

If the political setting for acid rain is a swamp, the economic setting might be the same swamp in a pea-soup fog. Numbers in the millions and billions of dollars are tossed about with abandon, some numbers apparently contradicting others. But two things are certain. First, acid-rain control in itself could become a significant industry. Scrubbers are enormously expensive to produce, often in the $100 million range, and they cost hundreds of thousands of dollars per year to operate. Second, the benefits of controlling acid rain certainly outweigh the costs, however enormous.

American industries spent an estimated $200 billion between 1971 and 1981 on clean air, and could be expected to spend another $200 billion between 1981 and 1988 under current law. (It is unclear how much of that total is for pollution "controls" such as tall chimneys.)

It is comparatively easy to calculate such control costs—to add up the purchase prices of pollution-control gadgets, the salaries of personnel to operate them, and the resources that must be purchased to keep them running—but determining benefits from pollution control, or, conversely, potential damage if pollution is *not* controlled, is considerably more complex. As with the health issue, assigning a monetary value to an entire aquatic ecosystem is one complication; another is to prove beyond a doubt that air pollution, or a given *type* of air pollution, was responsible for a given proportion of the damage.

Still, some attempts have been made to assess the costs of air pollution.

A 1979 study by the President's Council on Environmental Quality estimated the annual cost of acid-rain damage to architectural structures alone to be in excess of $2 billion.

In Europe, the Organization for Economic Cooperation and Development estimated in 1982 that a 50 percent reduction in sulfur dioxide would generate calculable benefits of from $1.78 billion to $15.54 billion per year.

In Canada, a government estimate suggested that the tourism and fisheries loss alone could reach $230 million annually. Much more ominous for Canada could be the prospect of damage to forests: a staggering 15 percent of total manufacturing employment in Canada is in the wood and pulp industries. In parts of northern Ontario, more than half of the labor force is employed in jobs related to the sport-fishing industry. The fishing-lodge industry alone is worth $50 million per year to Ontario. (Interestingly, 70 percent of tourists traveling to Ontario fishing lodges are Americans, most of them from the Midwest, where electricity is comparatively cheap and sulfur-dioxide pollution abundant.)

And various U.S. studies have suggested that savings of billions of dollars in health care and related costs would result from cuts in air pollution.

The list could continue, and if the United States were to invest the considerable research funds it would take to calculate such costs as fully as possible, the numbers would be awesome. To a certain extent, those numbers are important, but it would seem that the destruction of whole, vast parcels of northern wilderness would be a horrible "cost" in itself; and saving those parcels—the lakes, forests, and wildlife—a tremendous, incalculable "benefit."

Says scientist David W. Schindler, "There are those who insist that every living thing be reflected in terms of its economic value. Even these grasping individuals must be impressed by the damage of acidification."

Schindler might be wrong. In a September 1981 article, *Sports Illustrated* writer Robert Boyle reported that David Stockman, director of the Office of Management and Budget, once told a meeting of the National Association of Manufacturers, "I kept reading these stories that there are 170 lakes dead in New York that will no longer carry any fish or aquatic life. And it occurred to me to question . . . well, how much are the fish worth in the 170 lakes that account for four percent of the lake area of New York. And does it make sense to spend billions of dollars controlling emissions from sources in Ohio and elsewhere if you're talking about very marginal volume of dollar value, either in recreational terms or commercial terms?"

An EPA research scientist told me of an incident in which an electric-power company official expressed a similar sentiment. The scientist had just completed a study indicating that the hot-water discharge from a proposed nuclear power plant would severely damage the aquatic ecosystem directly downstream.

"So what does that mean?" the official asked. "That we'll kill some fish?"

"That," said the scientist, "and more."

"Well," said the utility official. "How much is a fish *worth*? We'll buy it! How much are a thousand fish worth? We'll buy 'em all!"

Perhaps what is really at the heart of the cost-benefit issue is an unfortunate but long-standing assumption that the atmosphere is a free commodity. The fundamental laws of physics state that matter is *conserved;* thus, injecting waste into the atmosphere does absolutely nothing to get rid of that matter; the matter will indeed remain. To convert sulfur-laden coal and some mineral ores into useful products like energy and metals, sulfur *must* be removed. And for centuries, the atmosphere has served as a convenient, "free" dump for that sulfur—no disposal fee, no operating cost. But it is reprehensible public policy and even worse economics, for it merely allows a handful of corporations and individuals to impose their waste-disposal costs on everyone else.

What sorts of political and technological action should be taken to stop acid rain, or even to gain reasonable control over it? A range of options is available. The Mitchell bill or a similar piece of legislation would be a sound beginning. In a comprehensive, surprisingly strong 1983 report on acid precipitation, the National Academy of Sciences stated that a 50 percent reduction in the number of hydrogen ions—the acidifying components of sulfur and nitrogen acids—would roughly halve the excess acidity in rain and thereby protect the most sensitive areas. These findings are based on studies from Scandinavia and North America that show that at acid-deposition rates of less than half a gram per square meter per year, even a fairly sensitive watershed can neutralize the relatively small amount of acid.

David Hawkins admits that the proposed 8-million-ton, 35 percent reduction envisioned in the Mitchell bill falls short of the levels called for in the National Academy report. It may not, he says, protect some of the most sensitive of North America's ecosystems. But, he says, it is at least a start, "a good first step in reducing that damage."

Beyond the Mitchell bill's proposed reductions, there are other possible legal remedies. Perhaps the most obvious would be strict enforcement of *existing* state emission limits, particularly the state secondary "welfare" standards, which are routinely ignored or waived. The EPA could act forcefully to disallow existing tall stacks as acceptable solutions to pollution. The Clean Air Act could be amended to allow no increase in sulfur-dioxide emissions when burners are converted from oil to coal. Congress could attempt to provide some sort of financial relief in the form of tax breaks or low-interest loans for aggressive industry pollution-control programs.

Perhaps most important of all, Congress and the EPA could examine seriously the possibility of pollution standards for acid deposition. (Present standards related to acid rain focus on gaseous sulfur dioxide and nitrogen oxides.) If such deposition standards were developed, they would at least allow the EPA to attempt to determine how much deposition a sensitive ecosystem could tolerate, and then to develop a regulatory scheme for reducing the rate of acid influx into that sensitive ecosystem.

In terms of technology, sulfur-dioxide scrubbers are not the only alternative, and in some cases probably not the best alternative. *Any* solution, however, will be predicated on the same simple principle. As Minneosta's acid-rain-research coordinator David Thornton puts it, "If you want to stop acid rain, you have to control emissions."

Some alternative pollution-control measures: mandatory changes in power-plant "dispatch" procedures, compelling utilities to use plants equipped with pollution-control devices as much as possible, and the older, dirtier plants only for peak loads; "coal washing," a process that literally washes 20 to 30 percent of the sulfur from high-sulfur eastern coal; and more reliance on other forms of electrical-energy production.

This raises the unavoidable issue of nuclear power as an alternative to burning coal. The Canadians boast of their heavy reliance, particularly in Ontario, on nuclear power as a "non-polluting" fuel. Indeed, nuclear power does not cause even a trace of acid precipitation, but its useful, practical role in an environmentally sound energy program is, to say the least, highly questionable. (Questionable, too, these days, in a financial sense. In 1982, a U.S. utility faced a twelve- to fifteen-year interval between filing the requisite applications to build a nuclear plant and the ultimate generation of electricity.) Despite any assurances to the contrary, too little is known about the long-term effects on human health of low-level radiation from nuclear power plants, and, further, there is, at this writing, still no convincingly safe way to dispose of the dangerous wastes produced in nuclear plants and no site to do so. (By 1982, almost all the nuclear plants in the U.S. were forced to resort to storing highly radioactive "spent fuel" wastes in on-site storage pools that were originally meant to store the waste for only a few weeks or months. To "expand" these pools, utilities were asking permission from regulatory agencies to place the spent fuel rods—the waste—closer and closer together in the pools. The capacity of the pools, however, is ultimately limited, since positioning the rods too closely together would initiate a nuclear chain reaction.)

There are other technological options that carry far less inherent risk, such as the so-called soft energy sources: solar, wind, geothermal power. On a practical scale, perhaps the soft energy sources could provide only token relief in the 1980s, but if a crash research program for such technology could receive even a significant fraction of the government funding devoted to bomber and missile-guidance systems, certainly some strides could be made—perhaps tremendous strides.

Further, as Eville Gorham points out, there is one acid-rain-abatement technique that is inexpensive, that carries with it no troublesome environmental consequences, and that is available immediately. This partial solution is energy conservation. President Reagan has repeatedly expressed contempt for the concept of energy conservation, tossing off the slogan "Conservation means being too hot in summer and too cold in winter." But, before the advent of Reagan, the Council on Environmental Quality looked at the potential benefits of energy-conservation practices (from insulating older houses and commercial buildings to requiring more efficient appliances) and a 1979 CEQ study noted, "The United States can do well, indeed prosper, on much less energy than has been commonly supposed." In fact, the CEQ said that with *existing* technology, "U.S. energy consumption in the year 2000 need not exceed current use by more than about 25 percent. And with a more determined effort . . . our total energy use could be held down to an increase of no more than about 10 to 15 percent."

Already, there is a new—albeit "hard"—technology under development that probably holds more immediate promise than any other: the "fluidized-bed reactor." This is a coal furnace into which powdered coal is injected onto a swirling, air-suspended bed of powdered limestone. The lime and sulfur react to form dry calcium sulfate and calcium-oxide particles that can be vacuumed out of the furnace. And, since fluidized-bed reactors operate at much lower temperatures than conventional boilers, the oxides of nitrogen do not tend to form as readily. The only hitch is that the furnace simply is not as efficient, yet, as a conventional boiler, although that day could come within this decade. In the meantime, sulfur-dioxide scrubbers will probably remain the best available technological fix to make conventional boilers operate cleanly.

Although the sulfur-dioxide scrubber works, it is a complex device that has its own set of problems. The only two scrubbers operating in the state of Ohio were installed in 1977 and 1978 at the Columbus and Southern Ohio Electric Company's plant in Conesville. During their first years of operation, the scrubbers functioned only 40 percent of the time because of a high rate of mechanical failure, although by 1981, they were functioning 80 percent of the time.

Not far across the Ohio border, at Shippingport, Pennsylvania, the 2,360-megawatt Bruce Mansfield power plant, a virtual cousin in power output to Gavin, consumes 24,000 tons of *high-sulfur* coal every day yet emits less than half of the EPA's old emissions standard of 1.2 million pounds of sulfur dioxide per million Btu's. The Mansfield plant is equipped with scrubbers—and good scrubbers at that. They extract 92.1 percent of the sulfur that pours out of the conventional boilers. To boost the benefit of the scrubbers even further, Mansfield's coal is washed before it reaches the boilers. One third of Mansfield's $1.3 billion construction cost was spent on pollution control, and a fourth or more of the man-hours at the plant are devoted to environmental efforts. According to the Cleveland *Plain-Dealer*, the average customer hooked into Mansfield's generators pays 7 percent extra for electricity—$1.40 on a $20 monthly electric bill.

But if scrubbers do present major performance and cost hurdles, at least one other major industrial country seems to be clearing them. In August 1981, the *Plain-Dealer* quoted Governor Rhodes—an avowed opponent of scrubbers—thus: "In Japan I asked one question, under my breath, 'What do you do when there's a conflict between [your] EPA and industry?' The answer was, 'We throw out the EPA.' "

What Rhodes apparently didn't realize was that there was a time, before 1970, when air pollution in Japan was so profoundly bad that a breath of oxygen could be bought from a coin-operated machine on a Tokyo street. But between 1970 and 1975, Japan reduced its sulfur emissions by 50 percent, even though its energy production increased by 120 percent. How? Largely by installing more than a thousand scrubbers throughout the country. Japan, in fact, has some of the most stringent sulfur-dioxide control laws in the world, and seems to have little trouble competing in the world manufacturing marketplace, even with U.S. companies that claim they can't afford to clean up their emissions.

Former CEQ chairman Gus Speth once compared efforts to control air pollution to the trials of Sisyphus, the cruel king of Corinth who was dispatched to Hades and sentenced for eternity to roll a huge stone to the top of a hill, only to have it roll back down as it neared the summit.

"It is important to reemphasize," said Speth, "that this latest chapter in the air-pollution drama might not have been as necessary if certain corporations had acted more responsibly by installing effective pollution-control measures years ago, instead of trying to get around ambient-air-quality standards by building taller smokestacks.

"When government is forced to act in response to this type of action, a kind of vicious cycle is created: government regulations tend to lead a corporation to abandon its own sense of responsibility and to adopt a philosophy that holds that whatever is not proscribed by law is permitted. This can lead to questionable corporate activities and, in turn, to greater regulation.

"The rock goes up the mountain; the rock rolls down the mountain. We can break this regulatory cycle only if business leaders, government officials, and the public work together to make social responsibility an everyday aspect of corporate decision making."

8

In terms of sheer territory, Canada stands to lose the most to acid rain. At some 3.85 million square miles, Canada is larger than the United States and second in area only to the USSR. The acid-sensitive Canadian Shield underlies most of the central and eastern portion of the vast country, stretching more than a million square miles from the rim of the Arctic Circle in the Northwest Territories to the Atlantic shoreline of Nova Scotia. A great deal of acid rain is falling on Canada, more than enough to destroy thousands of lakes and streams in coming decades, and much of it comes from the United States.

The greatest concentration of Canada's population is clustered in a narrow geographical band no more than several dozen miles wide and stretching along the lower Great Lakes from Windsor, Ontario, to Montreal, Quebec.

When a Canadian who lives in this belt buys or rents vacation property, chances are the property will be in the Muskoka or Haliburton "cottagers" regions that lie adjacent to each other, only a few hours north of Toronto, Canada's largest metropolis. On summer weekends, automobiles pack northbound four-lane Highway 400 bumper to bumper and when they reach Muskoka-Haliburton, around every turn and dip of the road lies another lovely lake rimmed with pine and birch and glaciated bedrock. There are the simpler cottages of Toronto's clerks and managers and the more luxurious summer digs of its social and financial movers and shakers; old and elegant tourist inns at lakeside interspersed with hundreds of live-bait shops, dozens of marinas, chichi boutiques, and woodsy saloons. There is a sizable population of year-round residents, some of them urban refugees but many of them born and raised here. A tiny but growing number are biologists and chemists who specialize in acid rain.

I visited the village of Dorset in the Muskoka Lakes region on a cool week in late November. There was snow in the forests already, and ice had formed around the shores of most of the lakes. I had come to visit a government laboratory and to see a professional fur trapper and official of the Ontario Trappers Association, John McClennen.

McClennen lives in a one-story log house that must have begun as a simple rectangular structure but has subsequently sprouted a couple of boxy additions. There is a sterotype for trappers, I suppose, as distinct as the cigar-chomping politician or the plaid-suited used-car salesman. As a stereotype, McClennen is a bust.

He met me at the door wearing one heavy white wool sock and one heavy brown sock. He'd just been out in his new car to check his trap lines, he said, and he ushered me into a warm, tidy parlor with wide picture window looking out upon a snowy hillside and a small creek below. There was an electric organ against one pine-paneled wall and bookcases, chock full, against another. McClennen insisted I take half his grilled-cheese sandwich and a strong cup of coffee. A kindly sort, he looked not like a trapper but like a scholar, or maybe a teacher, or maybe a librarian.

As it turns out, he is or has been all three: a university librarian for years in Toronto, until one day he wearied of "listening to the whistle," packed up, and moved back to the Muskoka Lakes, where he'd been born. Now he is an intermittent teacher of wildlife and forest resource management at a regional environment-education center, a player of the organ and the dobro, an occasional college student, and forest and wildlife manager of the thirty-five miles of publicly owned countryside for which he holds the sole commercial trapping license. From that land he takes beaver, muskrat, mink, otter, and other furbearers. In return for the sole rights to trap his section, McClennen is expected to ensure that there is no degradation of the wildlife population, of the woods, or of the water on those acres.

If John McClennen is outspoken about anything, it is the need to take only from nature what nature can be expected to regenerate. He produced charts showing how furbearer populations in Ontario had remained stable through recent decades and suggested that "despite a greatly reduced wilderness area, there are higher populations of these animals now than when the white man came to the New Land." He led me outside to a low shed to show me the soft pelts of beaver, otter, and muskrat hanging there, and he had harsh words for poachers who trap illegally during seasons when furbearer reproduction is important, and even harsher words for those who don't have the decency to use humane traps that kill an animal quickly. He showed me all this, he said, because his means of living make him acutely aware of changes in the routine patterns of nature.

It had been the tiny things that he'd begun to notice, alterations in nature any outsider might overlook.

"Something dangerous and subtle is amiss," he said.

My visit with McClennen came more than a year after Bill

Marleau had taken me to look at Woods Lake in the Adirondacks. What the Canadian trapper told me had elements of déjà vu.

He said that last time he saw frogs in the area was sometime around 1973. Muskrats used to leave clam shells piled up under his dock, but no more, for the clams seemed to be disappearing. In nearby Lake of Bays, where five years earlier he'd often catch a near-limit of trout early in the morning and be back in time for breakfast, trout are "no longer abundant enough" for him "to be bothered fishing at all." Bass populations had also dwindled, McClennen said. And he'd been noticing lately that when the faucet of his kitchen sink dripped, it left behind a strange, blue-green stain in the sink. He figured he knew enough chemistry to recognize the signature of copper sulfate, the likely result of copper's being leached from his plumbing system by sulfuric acid in the water supply. (There are villages in highly acidified regions of Scandinavia where the water supply runs blue-green for at least the first few seconds after the tap is turned on.)

McClennen's observations are not backed by any rigorous biological and chemical studies he has conducted, but even without government, university, or industry studies, he knows that his livelihood depends directly on the survival of Ontario lakes and their watersheds. More than 80 percent of the pelts taken in Ontario are from aquatic mammals—beaver, muskrat, otter—creatures that live in and near the lakes and streams and depend heavily or totally on food supplies that are aquatic.

Ultimately, he says, the issue comes down to something much more important than the survival of creatures of the Canadian woodlands. "Wildlife acts as a barometer of the health of the environment. If the wildlife begin to show signs of undue stress, there may or may not be time to correct the problems before they directly affect humans. Look, we live on a globe where 97 percent of the water supply is saline, and of the three percent that's left, a lot is tied up in glaciers at the North and South Pole. The portion that we're trying to protect—that people and animals can drink from—is infinitesimal. And in the Great Lakes basin, Americans and Canadians together probably hold the largest single supply of fresh water in the world. And what are we doing with it? We're bashing the hell out of it."

A few minutes' drive from John McClennen's house, the Ontario Ministry of the Environment maintains an acid-rain-research laboratory, a collection of large, modified mobile homes. Here, Canadians are measuring the acidity of precipitation as it falls into a fleet of collectors just outside the lab trailers and attempting to monitor the changes that this acidity is bringing to local lakes. Even more important, perhaps, the researchers based in Dorset are looking for a relationship between the dose of acidity and the responses of animals and plants in various waters. The idea is to find progressive early changes in ecosystems that might serve as signals of more wide-scale disruption to come.

European and North American scientists have accumulated a great number of data about the nature of already-acidified waters, but research into the early-warning signs of acidification remains inadequate. According to the lab's coordinator, biologist Ron Reid, "Everyone expects to come here and see a dead lake. And we're saying, 'No, we don't have any dead lakes yet, but we have some very, very sensitive lakes and maybe some that are pretty ill.' " And, said Reid, some of the early-warning signs of acidification are probably already appearing in some waters: subtle shifts in vegetation that resemble changes that have occurred in already-acid lakes in the nearby La Cloche Mountains.

"Certain species of phytoplankton that are indicative of

low-pH waters are appearing in certain lakes in this area," Reid said. "This summer we had those large balls of filamentous algae, for example, which tend to indicate more-acid lakes."

The Dorset lab's scientific staff had, in fact, coined its own name for the green, slimy, loosely packed balls of algae: "elephant snot."

There were other signs, too. The summer before my visit to Dorset, University of Toronto biologist Harold Harvey had run a controlled experiment in Plastic Lake, where John McClennen used to catch "nice size" rainbow trout. The pH of Plastic Lake was high enough during most of the year that fish should have been surviving, but there were no fish in the lake. So Harvey theorized that periodic surges of acidity might be plunging the lake into temporary acid shock from time to time. He guessed that the little bit of alkalinity left in the lake was inadequate to prevent a sudden acid surge from killing all the fish but still adequate to raise the pH after the shock was over. In other words, it appeared that Plastic Lake might be a body of water teetering on the verge of long-term and critical acidity.

When government scientists tipped off Harvey that a summer rainstorm had been even more acidic than usual, he packed equipment and rushed to Dorset. There he lowered a cage full of healthy fish into the waters of Plastic Lake and a control group of fish into another, better-buffered lake nearby. The control group survived; the fish in Plastic Lake did not.

At the lab, Reid had something else to show me. He produced an air-mass-trajectory map, one of many similar records of storm systems that had moved over Dorset. This map was a record of an August 1981 storm, the same weather system that had caused Harold Harvey to rush to Dorset with his cages of sacrificial fish, a storm during which the rain pH had dropped to an astonishingly low 3.2. The story the trajectory map told was that the system had first developed over eastern Tennessee, then had moved slowly over eastern Kentucky, over the Ohio River Valley, north toward Cleveland, and finally across Lake Erie, over Canada, and to the Muskoka Lakes. Here, the warm, muggy, polluted summer weather system had collided with cooler northern air, and acid-laden water had fallen from the skies and into the Dorset lab's collectors.

Repeatedly, air-trajectory analyses have shown similar patterns: when rain and snow systems move into Ontario from industrial regions of the United States, the pH of precipitation drops. It reminded me of what John McClennen had said just before I left: "Canada and the U.S. have the longest undefended border in the world, and we have to keep it that way, don't we?"

In 1972, when Sweden presented its landmark report on acid rain at the United Nations Conference on the Environment, it called acid rain "an undeclared act of war" between nations. "Act of war" might be a bit strong to characterize the reaction many Canadians have toward the United States on the acid-rain issue—but *only* a bit.

In October 1979, a joint United States and Canadian research report was released detailing the known and potential damage that acid rain could bring to both countries, including the suggestion that agricultural crops and forests could be in danger and the clear indication that each country was polluted both by itself and by its neighbor.

A few weeks later, a coalition of Canadian and U.S. environmental groups held a conference in Toronto. This "Action Seminar on Acid Precipitation," or ASAP, featured an extraordinary range of speakers: scientists from the fields of atmospheric chemistry, freshwater biology, and public health; government officials from both countries and from the United Nations; executives from Ontario's two largest corporate sulfur-dioxide polluters; and attorneys representing environmental organizations. But despite the depth and breadth of detail, it was a seminar designed for the general public and the nontechnical news media.

The ASAP conference suddenly attracted the attention of both the public at large (at least in Canada) and the media. Before long, articles with dramatic titles ("Death from the Skies") were appearing in magazines and newspapers. The cover of a Canadian magazine depicted a human hand dissolved to bone under the surface of a lake; sketches of skeletal fish were ubiquitous. A Minneapolis newspaper even carried a graphic representation of a prominent acid-rain researcher's face dissolving. Many of the leading acid-rain scientists, who had been plugging away quietly and steadily for years, were startled and appalled by all the sudden, and often sensationalistic, attention. But the issue had come of age at last.

Curiously, one of the major controversies that helped spawn the ASAP conference and accelerate the debate in North America was the proposed construction of the Atikokan power plant. Atikokan was the coal burner the Canadians planned to build and operate without benefit of sulfur-dioxide-control equipment, even though it would be located just north of the international Ontario/Minnesota canoe wilderness. By early in 1983, the Atikokan plant had been almost forgotten, but it deserves continuing attention for many reasons, not the least of which is that the Atikokan issue illustrates how sticky the international political web surrounding sulfur-dioxide control is.

The Canadian government is sincerely and profoundly concerned about acid rain. Canada's chief environmental official,

John Roberts, has called it the "single most important irritant" in relations between his country and the U.S. The provincial government of Ontario is sounding the alarm, and even the provincially controlled public utility that will operate the Atikokan plant admits that acid rain is a very real problem. Yet, in the mid-1980s, the Atikokan power plant will begin delivering electricity to western Ontario without scrubbing a single molecule of sulfur dioxide from its coal furnaces.

In the United States, the greatest contributors to acid rain are the so-called old power plants that do not have to meet stringent standards for new sources of sulfur dioxide. Since Atikokan is a new plant, scrubbers would have been required if it had been built a few miles south, in Minnesota.

Is it that the Canadians are so utterly blind that they cannot see the hypocrisy in demanding that the U.S. lower its emissions while Canada itself stubbornly refuses to match U.S. standards? Or could it be that the Canadians are using Atikokan as a bargaining chip in this international poker game? Maybe neither, for the web of circumstance is much, much more intricate than that.

Ever since the controversy began, some U.S. officials have insisted that it is unconscionable for Canada to demand further restriction from its neighbor in light of its own apparently lax standards. In 1979, EPA official Konrad Kleveno told the Toronto *Star*, "There's no question we're putting more acid into Canada than vice-versa. But there is also the matter of Canada getting its own house in order. There has to be a reciprocal action on the Canadian side if anyone expects the U.S. to clean up its sources. So far, we've seen no real action north of the border."

The Canadians counter that it is the U.S. position that is unconscionable. Canada is without question a major producer of acid rain. Although much of its electrical energy comes from hydroelectric sources and nuclear power plants, there is still a scattering of big coal-burning power plants, and moreover, nonferrous mining and smelting are major Canadian industries, and its smelters are monumental producers of sulfur dioxide—some 2 million metric tons per year, or about 40 percent of Canada's annual sulfur emissions. In fact, on a per capita basis, Canada is a greater emitter than the U.S.

But *total* Canadian emissions pale beside those from the United States.In Ohio alone, there are more than twenty coal-burning power plants, most equipped with no sulfur-control equipment at all. The eleven dirtiest of these plants pump out more sulfur dioxide than *all* the smelters and *all* the power plants in Ontario combined. Certainly, not all emissions from the United States wind up falling as acid on Canada. But one thing appears undeniable: if Canada reduced its sulfur-dioxide emissions to zero, acid rain would still be a problem there, for an estimated two thirds of the acidity falling on regions like Muskoka-Haliburton comes from the United States. The Muskoka Lakes lie only 250 miles upwind from Detroit, 150 miles upwind from Buffalo, and 280 miles upwind from Cleveland.

Further, say the Canadians, ambient, ground-level standards are roughly equivalent in the two countries. (However, the situation is complicated by the fact that in Canada, the provincial governments—not the federal government, as in the United States—have the only real authority to regulate emissions.)

And what all the bickering boils down to is this: U.S. emissions are projected to increase until almost the turn of the century; Canadian emissions will decrease.

Just before driving to Dorset, I interviewed Ron Taborek, Ontario Hydro's coordinator for acid-rain programs. "It shouldn't come as a surprise to people that if you dump thousands of millions of tons of sulfur into the air, it's going to come down in a way that is undesirable, considering the properties of the stuff," Taborek said. "We've made a corporate policy decision that we are going to be in the forefront of controlling the problem."

In fact, he said, Ontario Hydro was contemplating installing a scrubber system in its existing Lambton coal-fired plant in Courtright, Ontario, the fifth-largest sulfur-dioxide source in Canada and Ontario Hydro's largest. Utility officials in the United States have insisted repeatedly that such retrofitting of sulfur-removal equipment is so impractical that it is close to impossible. Taborek called the Lambton plant "a flagship operation to help prove it can be done."

But since it is much less expensive to outfit a new plant with scrubbers than to retrofit an old plant, why not install scrubbers at Atikokan? Taborek said scrubbers wouldn't do as much good at Atikokan, since the acid load downwind from the plant is currently much lighter than the load downwind from Lambton. Ontario Hydro's approach is to devote funds to controlling emissions that will do the most harm if uncontrolled.

"I think it's a very enlightened policy," he said. "Standard, good management says, 'Set your target and then get there in the most cost-effective manner,' in contrast to someone who says, 'put scrubbers on everything.' I think [the latter] is a thoughtless approach."

Atikokan, he insisted, is largely a symbolic problem—a problem of "how to talk to the Americans about cleaning up their emissions when [Canadians] have a coal plant just outside a wilderness reserve." He produced a map of North America dotted with pie charts. The tiny charts showed the estimated contributions of various Ontario Hydro coal-burning plants to rain acidity in various regions: Muskoka, northern Minnesota, the Adirondacks, and elsewhere. In each case, the predicted contribution to acidity from any one plant was a mere sliver of the pie.

Right there, Taborek had hit upon one of the great themes of the controversy. When I talked to officials in Ohio, they were quick to point out that acid rain was not a problem created solely by the state of Ohio. When I talked to officials in New York, they pointed to Ohio and virtually denied any major responsibility themselves. In Minnesota, Wayne Kaplan of Northern States Power noted that most of the state's problem was caused by plants outside the state and compared proposals to clamp down on his company's emissions to the futility of ridding one's lawn of dandelions when one's neighbors refuse to do the same.

From the very beginning, the same argument has come into play concerning Atikokan. After the initial alarm over the plant's construction was sounded by Minnesota in 1977, the EPA began a more detailed assessment of the plant's potential effects on sensitive ecosystems in northern Minnesota. Within months after EPA scientist Gary E. Glass began his study, two troubling facts became apparent: first, northern Minnesota waters were sensitive to acidification in precisely that same way as already-damaged lakes in Scandinavia; second, plenty of acid rain and acid snow was already pouring into those lakes. It was clear that a problem already existed in Minnesota, even without construction of the Atikokan plant.

Even though Ontario Hydro officials disagreed with some of the specific numbers in the EPA study, they triumphantly pointed to its findings as an indication that their own position on Atikokan was valid: amounts of acid rain added to the canoe country by the plant would be "small," and the plant's emissions would "not change the destiny of sensitive lakes in the area."

However, the EPA report said that Atikokan could increase

acidification in the area by as much as 15 percent. Ontario Hydro insisted that "somewhat less than 4 percent" would come from Atikokan, a number based in part on a reduction in generating capacity at the plant from 800 million watts to 400 million. (For the time being, it has been reduced even further, to 200 million watts; more capacity could be added later.) But to Ontario Hydro, the numbers weren't as important as the fact that Atikokan would not be *the* cause of acidification in the canoe country.

Every sulfur-dioxide generator in Canada and in the United States could make the same argument and could, in fact, prepare a pie-chart-studded map like Taborek's showing just how little any one source might be responsible for acidity in any given area. No single source would likely ever represent more than the tiniest of slivers of any of the pies.

Which changes not a thing. The assembled total of all the tiny slices still adds up to a mighty lethal pie. I told Taborek that if the map was intended as a reflection of Ontario Hydro's innocence, I wasn't impressed.

"Indeed," he said, "anybody could make the same point. So someone has to start."

It wasn't just idle talk. Although earlier in the debate some Canadian officials had expressed reluctance to move unilaterally, by 1982, provincial governments were clamping down hard on emitters. Ontario Hydro had been ordered to cut emissions by 43 percent, even with an expected steady annual increase in electrical demand. Similarly, large reductions had been ordered for the huge Inco Ltd. nickel smelter in Ontario and for other large smelting operations.

After years of merely exchanging rhetorical brickbats with the United States, Canada had begun to move on its own. In Ottawa, federal environmental official Raymond Robinson told me that the Canadian government has already resigned itself to the loss of some of the most sensitive lakes even with new stringent control measures.

> The kinds of reductions we contemplate will not stop the destruction of the most sensitive lakes, and will barely be sufficient to take pressure off the moderately sensitive areas. But there's no doubt, from what our understanding is of the processes in Scandinavia, that it will *slow* the rate of acidification. It's quite clear that what has happened in Scandinavia is more *years* of acidification. They've got far more acid lakes than we have, although their geology is no more sensitive than ours and the loadings [of acid] are no more than ours. There can only be one answer: it's a time factor. This is the hardest bit of science yet, to determine how far we are behind the Scandinavian experience. What we *can* say is that if you bring about the kinds of reductions we're talking about, there'd be no question that even in the most sensitive areas you could slow the process at which the acidity is using up the residual alkalinity.

Like other Canadian government and industry officials, Robinson insists that Canada alone cannot make a significant dent in its acid-rain problem without a transboundary air-pollution-control treaty with the United States.

"The heat is on Ottawa," he told an American audience at an air-pollution conference in Chicago in 1981. "As it is, the Canadian corporations requiring control can and do argue that there is no point in controlling our emissions if comparable reductions are not required of . . . American sources. While with Provincial support we have been so far able to overcome that argument, it has been at a price. That price is that we are working towards a meaningful air quality agreement with the United States."

Indeed, in fits and starts the United States and Canada *are* working in the general direction of an air-pollution treaty. In 1979, a Canadian official told a legislative committee in Ontario that such a treaty could be completed and in force by 1982. But 1982 came and went, and it was clear that any international treaty was far—perhaps very far—in the future.

Once again, the Atikokan power plant enters the story, for it was the Atikokan controversy that first induced members of the U.S. Congress to pressure the State Department for such a document in 1978. Actually, the notion of a transboundary air-pollution treaty had been tossed about for years, and there was precedent for it, since the two nations had already drafted and signed an international-water-quality treaty in 1972. The irony is that although the United States began pressing for an air-pollution treaty in 1978, by 1980, it was the Canadians who were insisting that an agreement was imperative. But no treaty exists.

In 1980, both countries signed a "memorandum of intent," calling for bilateral studies of acid rain, its environmental effects, the available alternatives to reduce sulfur and nitrogen emissions, and the legal and institutional modifications needed in both countries. The same memorandum called for both countries to "take interim actions under current authority to combat transboundary air pollution."

As far as the Canadians were concerned, the memorandum and the studies that followed were to lay the groundwork for a treaty. Joint scientific and technical "work groups" were formed to examine known data about the problem, assemble those data into coherent studies and reports, and make whatever recommendations would lead to controlling the problem.

Shortly after the Carter administration signed the memorandum, a Canadian Embassy official in Washington told me that the prognosis for action on acid rain was now very good. Once the crack scientific teams had released their reports, surely the United States government would see the justice of the Canadian position: emissions simply had to be reduced if important recreation and wilderness areas in the U.S. and in much of eastern Canada were to be protected.

The Carter administration had already gone on record as saying that acid rain was a serious problem. In his August 1979 environmental message to Congress, Carter himself had singled out acid rain as a serious environmental threat of global proportions, and had announced a ten-year program to assess the extent of the problem. There were, however, disturbing signals that the administration might not be listening to its own rhetoric. In 1980, Carter announced a plan to assist some fifty power plants in the Northeast in converting from oil to coal, in an attempt to reduce America's dependence on foreign oil. The plan would have provided $3.6 billion in direct aid to private utilities to ease the financial pain of conversion, but the proposal did not require the utilities to install sulfur-dioxide-control equipment at the converted plants, many of which would lie near the heart of the Northeast's most acid-sensitive ecosystems. U.S. utilities lobbied hard for the bill, eager to accept public funds to replace old oil-burning equipment with new coal-burning equipment. The bill passed in the Senate, but it failed in the House—not because of acid rain but because of its cost.

Despite any concerns the Canadians had about the so-called oil-backout bill and other Carter energy proposals, at least the rhetoric was there. In fact, EPA chief Douglas Costle, when asked about the oil-to-coal conversion proposal, told the Senate bluntly, "I think it's a bad idea." Costle and others at the EPA, as well as Carter's top advisers in the Council on Environmental Quality and environmental officials from several of the threatened states, continued to beat the drum. For the Canadians, at least there was hope.

Then, in November 1982, a new sort of environmental disaster hit Canada: the U.S. presidential election that swept Ronald Reagan and a flock of conservative U.S. Senators into office. The Canadians had already heard the extraordinary tales of Reagan's rhetorical mishaps regarding environmental issues—suggestions that trees and volcanoes were responsible for air pollution—and had heard him say that he would invite the polluting industries to Washington to rewrite the Clean Air Act. If there was a shadow of doubt about Reagan's intentions, it vanished when he chose as his Secretary of the Interior James Watt, a Wyoming attorney who had dedicated the previous half decade to fighting environmental and conservation laws and regulations. But despite Watt's key role in managing much of the nation's natural resources, he did not have direct control over regulations to abate air pollution. That control lay with another agency, the Environmental Protection Agency.

As administrator of the EPA, Reagan chose Anne Gorsuch, another former antiregulation zealot and a friend of Watt's. Gorsuch immediately changed the EPA position on acid rain. To her predecessor, Douglas Costle, acid rain had been an evident threat; the day Gorsuch took charge, not enough was known about acid rain to warrant any regulatory action.

Gorsuch let it be known that her EPA would support a relaxation of air-pollution standards, standards that already were far from adequate to prevent a predicted steady increase in both sulfur-dioxide and nitrogen-oxide emissions through at least 1995. In late 1981, Gorsuch issued her own acid-rain plan: more research. (Such research had already been ordered by Congress. In a memorandum leaked to the press, Gorsuch made her intentions clear: "Industrial groups will support this [more-research proposal] as they feel further regulatory measures would be premature given the present state of knowledge on the subject.'')

Needless to say, the Reagan administration proposal for acid rain called for no reduction in emissions.

In March 1981, Reagan visited Ottawa for talks with Prime Minister Pierre Trudeau. As Reagan appeared for his first public address to the Canadians from the steps of the Parliament building, a uniformed band began to play. From the crowd behind the band rose a huge banner that read ''Stop Acid Rain.'' Reagan said, ''New ways must be found to reinforce our special relationship. We live on the strongest, most prosperous continent on earth. But as we develop our resources, we must protect the environment around us. We will never shirk our responsibility to defend our way of life when it is threatened.''

But by October 1982, Gorsuch was telling a Pittsburgh symposium that controlling acid rain might somehow cause even worse environmental problems. ''We are finding a good deal of our data impressionistic, anecdotal, and contradictory,'' Gorsuch announced. She did not offer details as to how the halting of acid rain might cause even worse environmental problems. Rather, she quoted UN Ambassador Jeane Kirkpatrick: ''We have long believed that for every ill there is a cure. We should know by now that for every cure there is another ill.''

Meanwhile, the joint U.S.-Canadian effort to establish the basis for a treaty has continued, though not without further pain for the Canadians.

''It was not long after January 1981 [Reagan's inauguration], that we realized that the rules of the game were being changed even as the game was under way,'' says Ray Robinson.

Perhaps to "establish the mood," as Robinson puts it, the Reagan administration promptly altered the mission of the work group that had been assigned to develop pollution-control scenarios. A few months later, administration officials who had no expertness in acid rain began to tinker with the conclusions reached by the work groups. In one U.S. work group, two chairmen resigned in quick succession; U.S. scientists who complained about the suddenly altered procedure were reassigned to other tasks. The Reagan administration refused to provide travel funding for the scientists to travel to Canada, so most of the meetings had to be held south of the border.

Perhaps most distressing of all, prior to this manipulation from the administration, scientists from both countries had reached a consensus about a so-called target-loading: the degree of protection necessary to preserve most sensitive ecosystems. This numerical value would be the key factor in determining how much of the emissions needed to be controlled, but suddenly, in late 1982, the U.S. members of the work group switched course and refused to accept the figure.

Says Robinson, "What makes this situation particularly distressing, even absurd, is that at a twenty-three-nation conference in Stockholm on acid rain [in 1982], the United States representative endorsed a sulfur-loading target approved by the conference that is nearly twice as demanding."

For Canada, the projected damage from acid rain is staggering. Roughly eleven ounces of man-made acid are falling each year on every acre in southern Ontario. In that province, where there are an estimated 48,000 lakes sensitive to acidification, estimates of the number of lakes the province could lose by the year 2000 run up to *half* of that total.

In Quebec, more than 1,300 lakes are currently acid-stressed. Nine salmon rivers in Nova Scotia have been destroyed—the pH is below 4.7, and the fish are gone—and eleven other rivers are acutely threatened; their famous Atlantic-salmon runs could end within two decades. Acid rain is falling on Newfoundland and New Brunswick, both of which are extremely sensitive geologically, and on Saskatchewan and British Columbia, both of which appear to contain sensitive regions.

Ontario officals place an annual value of $600 million on the province's fisheries, a value of more than $1 billion on tourism. The loss to the Nova Scotian economy from the destruction of those nine salmon rivers is pegged at $300,000 per year.

Ray Robinson points to data indicating that acid rain can reduce the productivity of forest soil. The forest industry is Canada's largest, valued at about $20 billion a year, with annual exports of $12 billion, surpassing the *combined* revenues from agriculture, mining, fishing, and fossil fuels. One million Canadian jobs depend on the forest industry. Says Robinson, "Perhaps if you think in terms of your [United States] automotive industry, and the many sectors dependent on it, you can get a sense of how important the forest industry is for us in Canada. No Canadian government could view with equanimity a threat to this sector."

For the people who live in the Muskoka Lakes region, there are considerations beyond pure economics. I spent a long day driving about the Muskoka countryside with Sheilah Hatch, a District of Muskoka councillor (a position roughly equivalent to county commissioner in the United States), who wanted me to meet a few local people who were deeply concerned about acid rain. Most, she said, are "people who've lived up here their whole lives. They're not wealthy, not one of them. No health-food nuts or hobby farmers. Not the kind of people city people associate with 'environmentalists.' " Most of the summer residents, she added, are "more concerned about their designer sneakers than what's been happening to the lakes around here."

We stopped to visit Marguerite Stimpson, who had been born in Muskoka some seventy-one years before and who, as a blue-ribbon horticulturist, had recently been conducting some of her own informal acid-rain experiments. She had taken a butterfly begonia out of her living room and put it outdoors during the previous summer. When it rained, spots developed on the leaves. During three weeks of drought, the spots disappeared, only to reappear with the next rains. She showed me leaves of sugar maple, basswood, peony, phlox, flowering almond, birch, elm, plum, and apple. All were pocked and apparently burned in similar ways. She wasn't sure the damage was from acid rain, but the pocks appeared only when the leaves were rained on and didn't come from any insect or fungus she'd ever run into. There was one thing she knew for sure: "Frogs aren't singing to each other down in the bay anymore, and I can't find salamanders."

Several miles away, we dropped in on trout farmer Rolf Uhde, a tall, rangy, gray-bearded man who spoke of the changes he'd been seeing. Uhde raises rainbow trout, selling between 40,000 and 100,000 a month, and his ponds are naturally fed by a cold, clear stream that winds through his farm property. Recently there had been a heavy downpour—the same acidic rainstorm, it turns out, that brought Harold Harvey to Plastic Lake. The storm wiped out hundreds of Uhde's breeding stock. "People don't seem to realize that when I lose fish, all the other lakes are losing fish, too—while everyone's sleeping." He added, "You always take it for granted that the rain is clean."

At the Port Carling Ice Arena, we waited in the hallway

while Charlie Cameron, manager of the arena and chief of the local volunteer fire department, drove a small machine around and around, making fresh ice. In his office, later, he talked about catching female lake trout and finding their roe still in them, an ugly black. "It'd be rotten, dead right in their bodies. You'd catch them in the middle of winter and they'd still have spawn in them. And then in some of what used to be productive lakes, all you started to get was the bigger fish.

"But one guy," said Cameron, leaning his chair back against the cinderblock wall of his snug little office, "how's he going to get it across to the politicians?"

Charles Taylor, an official of the Ohio EPA—an organization that has been accused of dragging its feet on the entire air-pollution-control issue—told me once that most Ohioans did not see acid rain as a serious problem, if indeed they had ever heard of acid rain. He borrowed a term used often by James Rhodes, who was fond of railing against environmentalists. "The perception in Ohio is that there aren't many people affected, that it's the birds-and-berries people who are involved."

9

From the window of an Air Canada jet, I could see the forested rim of Lake Huron's Georgian Bay far below, craggy, glacier-etched, cut and nicked with little rocky coves and bays. The plane was full, which surprised me, since we were bound for Sudbury. From what I remembered of the city, it wasn't the sort of place one imagines anyone flying to—unless there's no choice. The DC-9 pulled away from the coast, and pine forest and lake flashed below. Suddenly we were decelerating, dropping.

Waiting at the gate was Bruce R. Dreisinger, tall, balding, and in white shirt and tie. Dreisinger's title is "Supervisor, Environmental Effects" for Inco Ltd., the largest producer of nickel in the world, one of the world's most prodigious producers of nonferrous metals in general, and according to *Fortune* magazine, one of the 500 largest companies outside the United States. Dreisinger had been working for Inco for a half-dozen years and before that was "Sulfur Fumes Arbitrator" for the Ontario Department of Energy and Resources Management, responsible for assessing damage from emissions from Inco's Sudbury operations, especially damage to vegetation. Now he does similar work for Inco itself.

It was a long drive into town, and along the way, Dreisinger told me that when the Ontario department he once worked for assigned him to Sudbury, he quailed at the thought of moving to such a place. But now he likes it well enough, his family settled in, his kids in school. I told him that I'd come to Sudbury in large part to see the Symbol. In the two days I was to spend with him, it was the only time I saw Dreisinger frown. Like anyone else who works for Inco, Dreisinger would rather I look at the big picture: at how Inco's products fit into the workings of the world in general; at how the labyrinthine mines and gargantuan smelters in Sudbury keep that city alive and contribute mightily to all of Canada's economy; at all the *good* Inco has done for the environment, especially the Sudbury environment, and especially in recent years. But he wasn't surprised at my comment.

"You won't be the first one," he said. "We've gotten pretty thick-skinned about that around here."

According to Dreisinger, ever since the explosion of Canadian news-media interest in acid rain in 1979, reporters, photographers, and television cameramen have come flooding into Sudbury to contemplate, write about, and photograph the Symbol, the structure that, at least in Canada, has become synonymous with acid rain itself.

"Oh, yes," Inco vice-president J. Stuart Warner had said to me earlier that day, in Toronto. "We're where we are today for symbolic reasons. Because of that stack, we're stuck with being a symbol, just like Marilyn Monroe was a symbol. And we won't get away from all of this until we become number two."

Inco is beleaguered, says Warner, because it had such a massive operation in one town—the world's largest nickel smelter, among other things. Everyone's pointing at Inco, he says, because the company just happens to put most of its emissions up just one smokestack.

"Maybe," he says, "if we put what we produce up two or three pipes instead of one, they'd ignore us."

Indeed, Inco's symbol—the Superstack, as it has come to be called—is only one chimney, only one pipe. But, oh, what a pipe.

Sudbury was prettier than I remembered it. There was a new enclosed shopping mall on the main street, a cluster of new government buildings, and, most surprising, an abundance of green grass and shrubbery. Dreisinger turned from the main street, Elm, onto Lorne Street, then around another short bend, and suddenly, there it was: the planet's largest chimney, dominating the western sky.

At 1,250 feet—nearly a quarter mile high—the Inco Superstack is not only the tallest smokestack on earth but one of the tallest structures of any kind, anywhere. (The Empire State building is the same height.) And from the mouth of the Superstack poured a billowing white plume of emissions that was equivalent to about 1 percent of the world's total output of sulfur dioxide from man-made sources. On the day of my visit, the Superstack was emitting sulfur dioxide at a rate of about 2,500 tons per day, but only because world demand for nickel was down at the time. Had the smelter been operating at full capacity, the sulfur emissions would have increased to about 3,400 daily tons. (Since my visit, limits imposed by the provincial government have reduced emissions to 1,950 tons daily.)

The stream of white gas spewing from the big chimney's tip usually would be expected to stretch into the distance in a tight, elongated cone, but today, an unusual weather pattern was flattening out the emission cloud, forcing it downward in a great, sweeping curl. This seemed to disconcert Dreisinger, who allowed that at the moment, the Superstack wasn't doing precisely what it had been designed to do; the visible plume normally would not have been grounding out, but a thermal inversion had moved over Sudbury, causing the plume to misbehave.

The Superstack was built in the early 1970s to prevent the smelter's pollution from settling to earth around Sudbury, to end an era when Sudbury looked like a vision of hell.

Although it lies in the middle of a vast, pinewoods region, the city of Sudbury was for decades a hallmark of industrial blight. Anyone passing through the city and its outskirts on the Trans-Canada Highway—as I did for the first time in the mid-1960s—was bound to be horrified by the landscape. I was in high school at the time, my family moving from northern Michigan to northern Maine. After riding through mile upon mile of northern lake and forest country, both my grandfather and I had dozed off in the back seat, only to have my father wake us to see a place that in his words, looked "like the moon." (He had no idea how apt the description was. At about the same time, the U.S. National Aeronautics and Space Administration—NASA—had requested and received permission from the Canadian government to test lunar-exploration craft and to train astronauts near Sudbury, because the ravished terrain was as close to moonscape as any in North America. Today, in fact, the winding streets of a newer Sudbury subdivision are called Galaxy Court, Telstar Avenue, Jupiter Street, Crater Crescent, and Moonrock Avenue.)

Back then, as I peered sleepily from a window of the family car, I saw acre upon acre of gray and black rock, littered here and there with white, twisted tree stumps that seemed bleached and dried like driftwood. Virtually everywhere on earth, some spore or seed will cling to the most hostile rock, find a way to attach itself, to persevere, to hang on, no matter how tenuously. But not here. Years of virtually uncontrolled air pollution from copper and nickel smelting had fumigated the countryside. There were long heaps of mine slag running for miles from the smelter. The Superstack wasn't yet there to reign over the region; there were other, shorter stacks at the smelter, at the company's iron-ore-recovery plant, at its Coniston smelter just southeast of town, and—as there is today—at the Falconbridge smelter east of the city. (With about two thousand employees, Falconbridge is the other major employer in Sudbury. But both its output of metals and its output of air pollution pale in comparison with its Brobdingnagian neighbor.) And mile upon mile of black, eerily barren rock, as if a giant disinfectant mop had somehow touched this patch of the earth and sanitized it of life.

During my months of research for this book, I heard repeatedly from Canadians—and from a few Americans who had seen the place—that they feared their own patch of the world might turn into "another Sudbury" because of the long-term effects of acid rain.

Yes, less than two centuries ago, Sudbury was a clean, green region of sparkling lakes and forests, much like the rest of the country nearby; and, yes, the devastation here had much to do with sulfur emissions. But Inco officials insist that Sudbury is *not* really a vision of the future for other areas on the Canadian Shield and in similar geological settings. They say the destruction here resulted from years of highly concentrated environmental abuse that no longer is being duplicated in Sudbury or anywhere else.

It may be too early to know how representative Sudbury might or might not be, in the long term, but for other reasons, a close look at Inco and its sprawling Sudbury operation is enlightening. Among nations, Canada may stand to lose the most to acid rain, yet here in its heartland, and sitting square atop the Shield, is the world's single greatest source of the problem. Whereas in the United States, coal-burning power plants are the major acid-rain culprit, Canada's nonferrous smelters emit *twice* as much sulfur dioxide as all other Canadian sources combined. As the country's greatest contributor, the Inco Superstack by itself pumps out some 20 percent of the nation's total output of the pollutant gas. Inco thus serves as a case study of Canada's struggle to deal with its own industrial sources of acid precipitation and, in a more general way, as an example of the problems facing many large industrial emitters everywhere in the world.

In the morning, Dreisinger and Inco environmental analyst Pat Bolger took me on a tour of the operation, really a host of related operations under one corporate umbrella. Miles of mines run under and around Sudbury, providing the raw ore that will in turn provide 5,000 tons of nickel daily, one fourth of the Free World's production. The Inco mills also produce daily 1,200 tons of copper concentrate and 2,500 tons of iron concentrate, along with smaller amounts of several other metals, for the few hundred square miles around Sudbury represent one of the most mineral-rich patches on the planet.

The two men drove me, first onto the vast plains of waste-rock tailings that surround the plant. Here, 1,300 of the 2,000 acres covered by waste rock from the milling process have been "vegetated" in the past decade—treated with lime and fertilizer and planted with fall rye and bluegrass, redtop, fescue, and timothy. Now some of the once-grim flats have become great, verdant meadows of waving grass, providing habitat for grasshoppers, spittle bugs, and ants. Deer mice and voles have moved in to become a small mammal base for this new ecosystem, encouraging predatory hawks, owls, and even foxes to join them. In the murky slurry ponds dug out among the tailings, ducks and geese have begun to make migratory stops and, in a few cases, have even begun to raise broods. "So we're not all bad," said Dreisinger with a grin.

We visited the site of the old O'Donnell "roast yard," several miles west of the smelter. The old yard looks like a long-abandoned airport runway; sparse grasses grew up to its edge, but the old yard itself was completely barren because of severe contamination by toxic metals.

The O'Donnell yard and others like it—they operated in the region through the last half of the nineteenth century and the first third of the twentieth—represent the first, and probably the worst, of a series of blows to the region's biota.

Here, nickel was smelted by the most primitive sort of pyrometallurgy. The Sudbury minerals are locked tight in a sulfur ore. To get at the valuable metal, the sulfur must be removed—not a complicated problem, since the sulfur will burn. From the surrounding forests were cut thousands of aspen trees, which were then piled four feet deep the length of the mile-and-a-quarter roast yard. The sulfur ores, dug by hand with pick and drill, were piled atop the logs, and the logs set afire to burn red hot for months at a time. The fires were hot enough to oxidize the sulfur in the ore into sulfur dioxide, a gas that would conveniently pour in thick, white, choking clouds from the smoldering roast beds, leaving nickel to be processed and sold.

Of course, the sulfur dioxide did not simply vanish. It rolled in great billows across the area's northwoods terrain, destroying first the most sensitive tree species, such as white pine and aspen, then the more resilient ones. With the tree layer gone, the shrub layer was stripped away, then the ground flora; then, with all the root systems gone, the soil's humus layer and finally the mineral soils below eroded away with the wind and the rain, leaving only bleak, gray, moonlike bedrock.

In 1930, the development of more sophisticated pyrometallurgical techniques brought the big smelters to town. The smelters were more efficient, more profitable, and, only by coincidence, cleaner—but not much.

For the next four decades, pollution output from Inco's big

smelter varied somewhat with production levels, but during the most productive times, when world demand for quality nickel was high, the Inco smelter was pumping almost seven thousand tons of sulfur dioxide into Sudbury's air each day—seven thousand tons that helped maintain the region as the barren, poisoned badlands that the roast yards had made.

Later, Dreisinger took me to see the smelter itself—a massive, dark, dirty structure several blocks long. As J. Stuart Warner, an Inco vice-president, had described the smelter: "It's really quite exciting—glowing red and white, hot and bubbling, and so on—like something out of the Middle Ages." Inside, great buckets of molten ore rode tracks the length of the building. Workers in hardhats moved about in the hot, thick air. Dreisinger told me that employees consider themselves lucky to be working here during Sudbury's cold winters.

If we had reached into one of the giant conveyors that carry Sudbury ore to the smelter and examined that sample of ore under a microscope, we would have seen a field of smoke-gray rock, flecked here and there with blue, yellow, and pink minerals. The blue is a copper mineral called "chalcoprite"; the yellow the nickel ore "pentlanite"; and the pink an iron mineral called "pyrrhotite."

Chalcoprite separates quite easily from the rest of the rock, but the nickel and the low-grade iron mineral are closely bound together in the sulfide ore. In a proportional sense, that pyrrhotite contains enormous amounts of sulfur compared with small bits of valuable nickel. Inco's scientists have reasoned for decades that the best, most cost-efficient way to control sulfur emissions is to remove as much of that sulfur-laden pyrrhotite (to "reject" it, as they put it) *before* the ore reaches the smelter.

Says J. Stuart Warner, "If we err on one side in the separation, we have an extra amount of sulfur going into the smelter. If we err on the other side, we are throwing unconscionable amounts of nickel away, high-grading the ore body and shortening the life of that deposit, which is also a serious social problem. So we try hard to optimize the separation.''

Since the 1950s, Inco's crack metallurgists have been progressively more successful at ''rejecting'' more and more pyrrhotite. By 1953, Inco was separating about 10 percent of the pyrrhotite, by 1963 about 35 percent, and by 1972 close to half. The company points out that its sulfur-dioxide emissions have therefore decreased by about half.

Still, the ore that does pour into the smelters contains about forty times more sulfur than nickel. And what happens in the smelters is not much different from what happened in the old, smoldering roast yards. It is chemistry sledgehammer style. The mineral ore is tapped out of the bottom of vats of hot, molten rock and injected into a roaring furnace where, as before, the solid sulfur is conveniently oxidized into a gas. The nickel remains behind—5,000 tons of nickel daily to be shipped around the world to become strategic parts of jet aircraft and bridges, alloy metals for our Fords and Chevys, and vital ingredients in the manufacture of stainless steel.

The waste rock removed in the concentrators is shipped to those vast plains of tailings to be revegetated and to provide habitat for voles and foxes. And the sulfur is shipped away, too, right up the big chimney. Since the operation began, Inco's refuse dump has been the cool northern air of Ontario, a cheap place to dispose of tons of unwanted sulfur.

Short of finding a way to reject more pyrrhotite before it can become part of the smelting process, Inco has only two further pollution-control choices. It can find a way to alter the entire smelting process so that sulfur is not converted to a pollutant gas, or it can find a way to capture the pollutant gas and convert it to a safely disposable by-product or, even better, a useful resource.

In the 1970s, Inco invested some $14 million in a nine-year project to heat the ore in a huge "pressure cooker" that would have removed most of the sulfur as a solid. The process change would have reduced the company's emissions to less than 10 percent of its former 7,600 tons per day, but the experimental project was ultimately abandoned because of high projected costs—more than $300 million—and operational complexity.

But in some of its other operations, Inco has had success with the other option—removing air pollution from the emission stream. At its nearby iron-ore processing and recovery plant, Inco produces as a by-product a sulfur-rich gas that can in turn be converted profitably into sulfuric acid and sold to a wide range of industrial customers. This or other existing technology could be used to control the emissions from the nickel smelter. Besides converting waste gas to sulfuric acid, the company might employ cleaner hydrometallurgical techniques, more-efficient "flash" furnaces, or scrubbers to filter the gas stream. But, says Inco, the costs would be enormous, costs that could reduce stockholders' profits, weaken Canada's trade balance, and put thousands of Sudburians out of work.

In 1970, the province of Ontario issued a "control order" to compel Inco to clean up the Sudbury operation. It was a two-pronged mandate. First, Inco was required to reduce sulfur-dioxide emissions from 7,200 tons per day in 1970 to 4,400 tons per day by 1975, to 3,600 tons per day by 1976, and, by December 1 of 1978, to 750 tons per day. Because of its successes with pyrrhotite rejection, Inco made the first two deadlines with months to spare, but the final limit—an almost 90 percent reduction in emissions—was never met. Too expensive, said the company.

But the actual emissions reduction was only half of the "control" effort. The other requirement was to abate the effect of those emissions. Provincial environmental officials were already taking heat from citizens and union officials about the unhealthy air of Sudbury, and there was little question that the air would *remain* profoundly unhealthy—that it would remain toxic to the region's vegetation—even at the reduced emissions mandated by the order. The province could have required an even faster emissions cleanup, but the company again responded that this would be too expensive.

So Inco and Ontario came to the only reasonable solution, the same solution employed at hundreds of other large industrial sites around the world: build a high stack. In this case, the highest stack ever.

On July 26, 1971, Inco official R. R. Saddington told the Toronto *Globe and Mail*, "Fears have been expressed that the tall stack will simply spread pollution over a wider area and put more poisonous sulphur dioxide in the atmosphere. In fact, it will do neither. There is widespread misunderstanding of the nature of sulphur dioxide. On a worldwide basis, a full 80 percent [!] of the sulphur dioxide in the atmosphere comes from organic decay. About 14 percent of it can be attributed to the burning of fossil fuels and about six percent to smelting operations."

He went on to say, "Sulphur dioxide survives only about four days in the lower atmosphere. It does not accumulate in the air as a poisonous layer in the earth's atmosphere. Therefore, the problem is not so much the volume disseminated from a stack but the ground level concentrations."

Saddington was right. There was "widespread misunderstanding" of sulfur dioxide, much of it on the part of those responsible for resolving pollution problems. Charts prepared by Inco show results of sulfur-dioxide dispersion at distances of up to a hundred miles from the giant smelter before and after the behemoth stack went into operation. The graphic record of close-proximity pollution for those months looks like a precipitous ski slope, with ground-level-pollutant concentrations falling abruptly as the Superstack comes into operation.

Local air quality improved vastly. Sudburians—like residents of other cities where high stacks had been added—could see and smell the difference. As an Inco publication puts it, "By the mid-1970s it was apparent that these measures . . . were working well and having the desired environmental effect. As Inco's [program] was operated with increasing skill and sophistication, the average and episodic ground level concentrations of sulfur dioxide in the Sudbury basin decreased. The vegetation began to stage a comeback. The qualitative observations of Sudbury's inhabitants agreed with the quantitative measurements of the experts—the problem of SO_2 emissions in the immediate area was under control."

Except for one hitch.

My motel room on the road between Sudbury proper and the smelter at Copper Cliff did little to promote the notion of either a "qualitative" or a "quantitative" improvement. Directly outside my window and built into the piles of waste slag and tailings was the "Big Nickel" tourist attraction, where one could have the "thrill of a lifetime" by visiting this replica of a nickel mine. And across the waste-rock plain, beyond the tourist attraction, was the Superstack itself. From this vantage point, it seemed astonishing that only a relative handful of academics such as Eville Gorham (who conducted studies on both acid and metal poisoning of Sudbury-area lakes after his return from England) would wonder about the real fate of those 4.5 million cubic feet of polluted gas that belch out of the chimney each minute.

With that view of the Superstack outside my window, I

flipped on the television for the evening news. As the picture was coming to life, a news announcer was saying something like ". . . and according to officials in Ottawa, Inco could reduce its emissions of sulfur dioxide by one thousand tons per day without any problem."

Suddenly, on the set, there was almost precisely the same view of the Superstack that I had from my room, as if the camera were on the roof overhead. Inco is, after all, big news in Canada. But considering the announcer's choice of words—"without any problem"—and considering that it was another in a long series of recent stories about Inco and the big chimney, and considering that the story's tone once again cast Inco as the bad guy in the acid-rain story, I had to imagine J. Stuart Warner, the Inco vice-president responsible for environmental control, sitting somewhere in his living room or den and scowling at the CBC announcer. To put it mildly, the publicity has been getting on Warner's nerves.

Warner's spacious office on the forty-fourth floor of a Toronto tower looks down into the great canyons of glass in the central business district of Canada's largest city. There was a potted tree in one corner of the office and a wide expanse of desk piled high with fat folders and reports. Warner is fortyish, holds a doctorate in engineering science, has taught thermodynamics and kinetics at Columbia University, and is coinventor of a complex pyrometallurgical process. For years he ran Inco's major research laboratory.

"There's no denying," he said, "that we're a large source of sulfur dioxide—in fact the largest point source in the world. But there's a certain air of villainy attached to it, and *that's* what sets our teeth on edge. Maybe we *are* a big operation, and maybe we do put out a lot of sulfur dioxide. But *so what*? There's a very fundamental principle of *justice* being overlooked here."

Inco, he insists, is only minimally responsible for any acid-rain damage in Canada. But, he says, because Canadian politicians and regulatory bureaucrats have so far been unable to handle the *real* source of the problem, they pick on Inco. "Show biz," he calls it. "And window dressing."

The real issue, according to Warner, is that "the United States is a very large net exporter of air-pollution damage to Canada." Warner reached beside his desk and produced a fat folder of newspaper clippings about acid rain and Inco, and banged the folder down on his desk. "And this is just the recent stuff. You really get leery after a while, when you're always made out to be the fool. Or maybe it's the knave. I don't know which is worse, but it's a boring role, I'll tell you. Everybody thinks so badly of you—what the hell—you just stop caring what they think anymore."

Unlike many of his counterparts in the United States, Warner doesn't deny for an instant that industrial pollution causes acid rain, or even that Inco is a significant contributor. What infuriates him is that the Canadian government and the Canadian news media continue to behave as if Inco's Superstack is *the* cause of acid rain. "Wouldn't it be wonderful if we were? God, schoolchildren would kick in to help shut us down." If the Superstack were the cause of acid rain, controlling the problem would merely be a matter of shutting down the eighth-largest corporation in Canada—putting 14,000 workers on the dole—and strangling the world supply of nickel. Phenomenal disruption, yes, but in one swoop, Canada could save millions of acres of its now-threatened territory, and without any messy international politicking or tough energy policymaking.

But it is far from that simple. During a routine shutdown of the Sudbury smelter in the summer of 1978, and again in the midst of an eight-and-a-half month labor strike in 1979, researchers for the Ontario Ministry of Environment conducted extensive monitoring of rainfall in and around Sudbury. The scientists had established that acid rain was falling in the area, falling not only in Sudbury but also scores of miles away, in the heart of the province's top vacation regions, the districts of Muskoka and Haliburton. With Inco shut down, the researchers finally had an opportunity to determine the degree to which the Superstack was responsible for acid fallout in northern Ontario.

Warner characterizes the provincial authorities as thinking thusly: "Boy, have we got 'em now. This is our chance. We're going to run around and sample the rain and what we're going to find is rain pure as drinking water when Inco's not operating." Warner said, "Know what they found? No change. *Nooo change* in the acidity of the precipitation. None whatever."

Warner may have been exaggerating a bit. In a corporate publication featuring an interview with Warner, he said that the data showed a 10 to 20 percent Inco contribution to rain acidity in the immediate Sudbury area and "very little" contribution in the Muskoka-Haliburton region. If those government figures are accurate, they bear out Warner's claim that Inco is not the primary cause of acid rain in Ontario but one among hundreds of sources in the northeastern quadrant of North America.

Warner said, "A lot of our guys, when they saw the results, said, 'Hey, we're off the hook. We don't do anything.' I said, 'Wait a minute. That's baloney. We can't say *that*. But what it does show, beyond any shadow of a doubt, is that we are not *the* cause of acid rain, as some idiots have said, and that we are not the *major* cause of acid rain. Our contribution to the acidity of precipitation in the area is competely overwhelmed by what comes in from long-range transport.' "

He rummaged in a file folder and produced a map of North America. There, he said, pointing to Detroit, Chicago, and points south and east, are the sources about which Canada should really be concerned. He suggested that the types of air masses that tend to form over the warmer, muggier regions of the midwestern United States may be more conducive to acid formation, and, further, that the travel time from the industrial Midwest to sensitive regions in Canada may be key. By the time polluted air masses reach Canada, he said, "the clouds are ripe. The sulfur dioxide and nitrogen oxides have been oxidized by then and are ready to be brought down by precipitation."

Like the Canadian government, Warner insists that what is really needed is a treaty between the United States and Canada. He envisions a treaty calling for tight controls on all new sources of acid precursors, a major international effort to find ways to control the oxides of nitrogen (neither of these would affect the existing Sudbury smelters), and a close examination of all existing sources to see if improved pollution abatement is "technically and economically feasible." (If a project is not economically feasible, then it isn't really technically feasible, Warner says.)

Unlike most of his counterparts in the United States, however, Warner finds fault with arguments that too little is known about acid rain, that not enough scientific information is available to require further pollution controls. In fact, in a 1982 letter to the U.S. Congress, Warner stated, "I have seen no scientific argument advocating increased emissions of SO_2 and NO_x [sulfur dioxide and nitrogen oxides]. I have seen many arguments that even present levels of emissions are too high. The minimum response I expect from someone who places great weight on science would be to ensure that *no increase* in emissions would be tolerated. This is not what we are seeing in the U.S."

But Warner scoffs at Canada's potential for success at the international negotiating table. "The EPA," he says, "is trying to sort out its own problems, and the last thing they want is a few Canadian diplomats running around down there in Washington trying to negotiate a treaty. So I think they give them a few chores: 'Well, you go home and take care of this.' And that'll keep them out of town for six months or so."

Those chores, he's certain, consist largely of instructions to clamp down harder on Inco and a few other large Canadian industries and utilities, a process that caused the Ontario government to order Inco to reduce Superstack emissions to less than 2,000 tons per day by 1983. Warner compares the order to trying to solve an entire town's bad-breath problem by "finding the tallest guy in town and pulling all his teeth."

He adds, "With a debate going on about whether the U.S. should keep emissions at the *same* level when the *same* level is too much—when it's causing a *lot* of damage in Canada—it looks like the political process will sacrifice the environment to the coal interests in exchange for votes in those states. This is what's so discouraging to us."

Warner insisted that Inco is profoundly committed to further sulfur-dioxide reductions. Although the company protested the provincial control orders to keep emissions below 2,000 tons per day because this would merely force Inco to limit production, he vowed that the company would leap at any chance to maintain production levels while controlling sulfur emissions.

"When we came upon a way to do the extra pyrrhotite rejection, we did it. We did it right then. We did it well ahead of schedule. We didn't wait for the government to say, 'Okay, tomorrow you have to crank it down.' And as soon as we find the next thing to do, we'll do it *right then*. Whether the government orders us to or not."

I pointed out that the company could still operate at maximum capacity and reduce its emissions by anywhere from 50 to 95 percent using already-available scrubber technology.

Warner looked at me as though I had truly lost my mind. "Ahhhh," he said, throwing up his hands. "Wet scrubbing is anathema. Look, we're resource people. Scrubbing takes one resource, limestone, and another potential resource, sulfur dioxide, and turns them into another environmental *problem*."

Scrubbers produce a calcium-sulfate sludge that, in most North American cases where scrubbers are operating, must be dumped and stored indefinitely in leak-proof lagoons. Calcium sulfate, however, is gypsum, and the Japanese consistently use the material from *their* sulfur scrubbers for the manufacture of such products as wallboard.

"But," Warner counters, "in Sudbury, we're not talking about a little pile of sludge. We're talking about a huge, huge pile."

I pointed out that a number of Canadian officials and environmentalists seem convinced that the facility could operate at full capacity at greatly reduced rates of sulfur-dioxide output with new, high-technology, clean-burning smelters. He replied that the company had in fact invested millions of dollars in the past decade in research to determine if the company could install super-clean processes. Those studies did prove that such processes were possible, but at costs in the hundreds of millions of dollars. But, he reiterated, Inco remains committed to further reductions in the future.

"Depending on cost," I said.

A cloud passed over Warner's face. "Well, that's what I mean! I'm not talking about *making* money. There's no way we're going to make money on pollution abatement. But this is the sort of thing we've been trying to do all along, to find ways to reduce our emissions that make sense for us as well as for the environment."

And how would Warner respond to those who say the investment in pollution control pales in light of profits that run into the hundreds of millions of dollars annually or beside the staggering value of Inco's mineral rights to the still-rich Sudbury ore deposits, which will produce large corporate profits for decades? Or to those who point out that Inco, like industries everywhere, still operates on the presumption that the atmospheric environment should be a free waste dump, thus allowing the corporation to decrease its costs and increase its profits at the expense of everyone else?

"Companies," he said, tapping the piles of memoranda and reports on his desk, "are just paper, as you can see here. We just pass things through. We don't make nickel for ourselves, and the power companies don't generate electricity for themselves. We do it for people out there. And the people pay. They always have and they always will. You know, the environmentalist who goes around selling this garbage about 'the polluter must pay' is deluding himself and others. The people *always* pay. They pay in products. Or they pay in taxes. Or they pay in environmental quality.

"But, look, we don't go poor-mouthing or worse, doing a Chrysler [referring to Chrysler Motor Company's efforts in the early 1980s to find government financial support to keep the struggling company afloat], saying 'Hey, we want to be good guys, but we've bitten off more than we can chew. Now help us.'

"We don't *do* that sort of thing. It is going to take process improvements. But you can't always find process improvements tomorrow."

Warner snapped his fingers high in the air, flamenco style. "Do this today." Snap. "Do this tomorrow." Snap.

My last morning in Sudbury, I contracted cab driver Curry Armstrong to take me out early in the morning to see the local sights free of corporate guides. We drove to the still-devastated terrain east of town near the defunct Coniston smelters, and Armstrong waited on the shoulder of the highway while I climbed a bare road cut of near-black bedrock and onto a vast plain of desolation. Someone had tossed an old stained mattress up here and a couple of cases of beer bottles. There were a few contorted, stunted shrubs holding on to bits of soil in the cracks and crevices, and even a few swatches of grass. But here it was hilly, and, as on many rolling acres around Sudbury, most of the soil had washed away long ago.

I snapped some photos of the ruination, and we drove back into town. According to Warner and Inco's public-relations material, the company has changed its tune. In the days of the old roast beds, and even in the early decades of the big smelters, destruction and desolation were simply part of doing business; a wholly different ethic was at work. Maximum profit was the objective, and if the forests and lakes around here were wiped out, there were many more forests and lakes elsewhere.

"I've been around here since 'fifty-two," Curry Armstrong said over his shoulder. "You couldn't even have a garden back then. You'd get it going all right, but then the wind would change direction, and—poof!—no more garden."

At least the Superstack has changed *that*. Curry Armstrong has a garden now. The company had given me a brochure detailing its local improvement efforts: "To the tourists and travellers programmed to the moonscape idea of Sudbury by the national media . . . the question must arise . . . whether the reporters were objective in their observations. Local residents comment favourably on the many improvements and make use of the recreational facilities provided."

It depends on where you look. Inco has indeed brought in truckloads of soil, lime, grass seed, fertilizer, sod, and shrubs, but much of the carefully orchestrated "improvement'' has occurred only in areas directly adjoining or otherwise visible from public roads. The company has put up ''aesthetic barriers'' to block from public view some of the operation's least inviting scenes.

That evening, there was some sort of street carnival downtown, with merchants hawking sneakers, T-shirts, socks, records, and books at sidewalk stands, and a drum-and-bugle band called the Northern Brass rumteetumming on the corner of Elm and Durham. It seemed a thriving city, this town that Big Nickel built, especially with its new government high-rise office building and in-town shopping center.

Presiding over it all, with the sun setting behind it in a stunning blaze of pink, red, and orange, was the Superstack. I tried at the sidewalk sale and in the drugstores and shops to find a postcard of the town's most prominent feature, but there were cards showing only the paper birches and the flowers in Simon Lake Park and Bell Park.

10

Is it possible that most of what has been said by dozens of prominent scientists, concerned environmentalists, and the news media is really an illusion?

The weight of scientific opinion clearly indicates otherwise. Not one major report has suggested that acid rain does not exist or that it is not a problem in sensitive areas. Yet there are those who continue to insist that much of what has been said about acid rain is the result of inadequate research, poor research, or hysterical media attention to existing research. These opponents of immediate regulatory action suggest that the alarm about acid rain simply is not warranted. Writes Alan W. Katzenstein, a public-relations consultant to the electric-utility lobby, the mere term "acid rain" is one that "conjures up images of destruction far beyond anything documented to date."

Was the editorial in the *Wall Street Journal*, on September 17, 1982, entitled "Gitche Gumee's pH," correct when it said that research really "points more and more toward the theory that nature, not industry, is the source of acid rain" and that "startling information is coming to light that suggests that acid rain has very little, if any, connection with the dying lakes in Canada?"

These appear to be astonishing statements in light of evidence from a number of major scientific reports from Norway, Sweden, a team of Canadian and U.S. scientists, a similar team commissioned by the Environmental Protection Agency, and a panel of the U.S. National Academy of Science. In fact, the *Wall Street Journal* editorial is inaccurate. No such "startling information" exists, although incomplete and misleading public-relations documents peddled by the industries' lobby might tend to lead a sympathetic editorial writer to those conclusions.

If acid rain is causing real damage, the only credible argument for regulatory inaction is that action would severely disrupt the North American economy. As Stuart Warner put it, companies do—and must—"just pass things through." If U.S. utilities are forced to clean emissions beyond present levels, the costs will without question be borne largely or wholly by consumers. Even if rate increases for pollution control amount to only a dollar per month per customer, those dollars might otherwise be invested in goods or services that create jobs and profits. Cheap electricity, after all, makes for less expensive consumer and industrial products that therefore will be more competitive in an international marketplace.

Of course, there are counter arguments. If the abatement costs calculated by the Congressional Office of Technology Assessment for the Mitchell bill are any indication (an average 1.9 percent increase in utility bills), the *degree* of any disruption is certainly minimal, especially in light of the steady increases energy companies are already passing along to consumers. Further, there seems little moral justification for consumers in a cheap-electricity state like Ohio to benefit from money-saving schemes that call for dumping pollution across state and national borders, the equivalent of my saving sewage disposal costs by pumping my toilet effluent into my neighbor's dining room.

Further, as environmentalists and many economists have pointed out, pollution control itself is a productive industrial activity—assuming, of course, that clean air is a valuable commodity. When millions of dollars are spent to build pollution-control equipment for a power plant, thousands of jobs are created in the industries that design and construct that equipment. To suggest that the clean-air industry is "unproductive" is tantamount to suggesting that the cold-air industry (refrigeration and air-conditioning) is unproductive—something the electrical utilities would be the very last to do.

I tend to see the environmentalists' argument as more reasoned. Presumably, there are those in the boardrooms and executive offices who could offer a more well-developed rebuttal, but I couldn't find them, although I found abundant comment on the acid-rain issue from the utility and coal industries.

The utilities, especially, have produced a cargo of brochures, booklets, and assorted handouts all directed to one purpose: to stall regulations on acid rain while more years are spent conducting research into the issue. The Edison Electric Institute, an industry lobbying and public-relations organization based in Washington, D.C., insists that its real mission is to "ascertain factual information, data and statistics related to the electric industry, and make them available to member companies, the public, and government representatives."

But the Edison Electric Institute also is the major producer and disseminator of antiregulation material on the topic of acid rain, developing a veritable blizzard of misleading statements, half-truths, and distortions blended with some accurate information. Two publications from the institute are entitled

"Before the Rainbow: What We Know About Acid Rain," edited and with an introductory chapter by Carolyn Curtis, and "An Updated Perspective on Acid Rain," by Alan Katzenstein. Curtis observes, "By definition, natural rain is somewhat acidic. . . . By all rights we should have been saying for years, 'It's acid raining outside,' or 'Take your umbrella. It's going to acid rain today.' This sounds preposterous, but it's true. Thus, our first understanding is that the strong verbal image, 'acid rain,' elicits more fear than it deserves."

Katzenstein attempts a similar semantic diversion. In a section entitled "Acid in Every-day Life," he writes, "Chemicals are part of every living plant and animal, and many of these chemicals are acidic. . . . In our vocabulary, 'acid' often implies strong actions, such as 'acid test,' or even unpleasantness, such as 'acid-tongued.' But acids are not necessarily undesirable, unpleasant, harsh or a hazard to life. . . . Taste is our most sensitive detector of the acidic nature of substances we contact in daily life, but it often deceives us. It is not surprising to find that tomatoes are acidic but most people are surprised to learn that a delicious pear can be more acidic than a tomato or that bananas and carrots are nearly as acidic. All of these have pH values well in the range of the rain that is the subject of scare headlines in the popular media."

Katzenstein also has a unique slant on "The Effects of Acidity": "This absence of verifiable damage in the plant world is not surprising. Variation in the environment is part of nature's plan; different varieties thrive in differing environments. Azaleas, rhododendrons and evergreens do better with high-acid fertilizers while grass, flowers, and vegetables call for low-acid nutrients. Lawns look better when they have been properly limed to raise the pH, but many yards have a satisfactory look even when people fail to check the soil pH or to correct for the acidity that has built up over time."

My own guess is that almost all readers can see through such blatant red herrings. Everyone to whom I've read Katzenstein's acid-banana classic has been able to appreciate it for what it's worth.

More insidious are arguments for inaction based on half-explained results of scientific research. The most frequently cited of such arguments involves a study in the Adirondacks—a study of Bill Marleau's acidified Woods Lake and two other alpine lakes, Panther and Sagamore. Called the "Integrated Lake Watershed Study," this investigation was funded by the Electric Power Research Institute, the research-funding arm of the electric utilities. Although there is wide agreement in the scientific community that EPRI's research is top-notch there is little question that EPRI management officials and the utilities themselves have used convenient bits and pieces of the accumulated data to help cloud the issue.

In 1980, Ralph Perhac, an EPRI official, told a Congressional committee, "In EPRI's lake acidification study, we have found three lakes in the Adirondack Mountains of New York State which have very different acidities, yet these lakes lie within a few miles of each other and chemistry of the rainfall is the same at all three. *Obviously, some factor other than precipitation is responsible for the acidity*." (Emphasis added.)

Perhac's statement is simply wrong; the three-lakes study showed nothing of the kind. It did show that lakes in low-alkaline watersheds are most susceptible. Researcher David Thornton states, "The total amount of soil in the three watersheds is different. And the rate of acidification clearly depends on the total amount of soil, along with the kind of soil, the amount that comes into contact with the rain, and the hydrological characteristics of the watershed. In fact, Woods Lake is the most acid lake and the one with the least soil in the watershed. The EPRI work is outstanding research and it is *proof* that acid rain is causing lake acidity."

But even though Perhac's statement was, at best, misleading, later in 1980, lobbyists for the coal industry were presenting the same argument virtually word for word to Congress, and a year later, Alan Katzenstein resurrected it in his "Updated Perspective" booklet. Katzenstein concluded that "the varied and complex interacting influences make it impossible to pinpoint any single factor as being the major determinant of lake acidification or of fish loss."

Those who suggest that something other than acid rain caused lake acidity fail to explain fully that the other "major determinants" they refer to are the geological and hydrological settings of the lake. An equivalent bit of illogic: my neighbor punches me in the nose; when I sue, he argues that to blame him—"a single factor"—is unfair and scientifically unsound, for my nose was in the direct path of his moving fist. Similarly, Katzenstein appears to be arguing that the lakes are to blame for being low alkaline and for lying directly in the path of the polluted air masses his employers help create.

The utilities and coal companies do, however, offer a solution: pouring lime into acidic lakes to raise the pH or into threatened lakes to maintain some measure of buffering capacity. After stating that, "So far, there is little evidence that real benefits would be realized" from installing scrubbers on Midwestern coal plants, Katzenstein notes, "The most promising strategy for helping lakes with low pH problems is to use lime or limestone to reduce the acidity and raise the pH."

In fact, researchers have experimentally limed lakes in Scandinavia, the Adirondacks, and Canada with varying degrees of success. In New York, the cost ranged between $30 and $300 per acre of lake at a rate of about one ton of agricultural lime per acre. But the measure is only temporary as long as acid rain continues to fall. In the most extensive liming experiment ever conducted, some 1,000 metric tons of lime were introduced into the entire watershed of the River Hogvadsan in Sweden over a four-year period in an effort to raise the pH to a level that would support salmon. But during acid onslaughts accompanying snowmelt and heavy rainfall, the experiment proved to be a failure.

Katzenstein states that "liming is a bargain when compared to the increased cost of electricity if retrofitting with scrubbers were to be imposed." Although he offers a figure of $4 million a year to raise the pH of 468 of the Adirondacks' most poorly buffered lakes, he does not estimate what it would cost to maintain the additional tens of thousands of lakes and streams across the Canadian Shield. Further, he fails to mention that raising the pH will not restore an acid ecosystem that has already been subjected to a related assault of mobilized aluminum and other toxic metals.

Ron Reid (no relation to his namesake at the Dorset laboratory) of the Federation of Ontario Naturalists says, "You can add lime to raise the pH but that doesn't necessarily bring the other chemicals back into balance. What you do is create pretty much of a chemical soup."

But there do remain some legitimate doubts. The utilities often point to the landmark data developed by Gene Likens of Cornell, in the early to mid-1970s, that show a trend toward increasing rain acidity. But the utilities point out that Likens was working from a smattering of historical data collected by varying procedures and at different stations, and thus state that the evidence is inconclusive that the rain is becoming *more* acidic. To some degree, this is something of a red herring, too, for it matters little whether there is an upward trend if the rain is already too acidic for sensitive waters. Similarly, while the utilities admit that some lakes are clearly too acid for fish to reproduce or survive, they suggest that since good historical-trend data for the *lakes* do not exist, few solid conclusions really can be drawn. Further, they argue that scientists have not finally proved, beyond doubt, that pollutants can ride the winds hundreds of miles or that power-plant emissions are really the culprit.

Eville Gorham agrees that better trend data are needed. "But," he says, "what [the utilities] do is fix on those cases where evidence is doubtful and ignore the weight of much more solid evidence. My feeling is that the circumstantial case is now so solid that we have to do something. The only question is how to go about it. I think we've just got to start rolling back the emissions and then look empirically at the result. If we had waited for a chain of cause and effect to be established perfectly after the London Smog, nothing would have been done about it *yet*. [In London] they *still* haven't tied everything down to a faretheewell. But they had a good, straight, circumstantial case. They acted on it. And they were right."

Reid agrees: "Yes, there are lots of areas where the answers we have are somewhat tentative, even in terms of the kinds of effects it has on lakes or how you determine how many lakes are affected in the future. There's a lot of uncertainty on the wind patterns, the mechanisms in which acid rain is formed in the atmosphere. There's no question that there are a lot of unresolved factors concerning control technology, whether scrubbers are the best technology, whether a reduction of 50 percent or 75 percent will be enough. But we don't really have the luxury of waiting until we have all the answers.

"You're talking already about five years' construction time

to build and install scrubbers. If you add even five to seven years of research and another two or three years to write regulations and make all the related administrative decisions, you're suddenly talking about fifteen years before you even get the scrubbers in place. How many lakes do you lose along the way?''

Charles Taylor, Ohio EPA air-quality chief, says, ''What happened to those [acidified] lakes may be the result of a hundred years of human activity. Maybe drastically cutting emissions right away won't make that much of a difference. I don't think another three to five years is going to make much difference. I don't think any lakes are going to be wiped out that quickly. Well, maybe some of the most critical.''

However, the University of Maine's Steve Norton adds the best perspective: ''I would not like to see us run it as a long-term experiment and have it turn out that we were right.''

11

I drove with David Thornton along the shore of Kane Lake, a smallish body of water in northern Minnesota that lies well south of the canoe-country wilderness. It is not a particularly remarkable piece of surface water; the terrain surrounding it is scrubby forest, there are no encircling bluffs or pretty coves, and most of the cottages near it are little more than old fishing cabins. On the summer Sunday when Thornton and I visited, a few small motorboats were dawdling along; a larger boat towed a water-skier. There was no apparent public access, but we found a small stretch without a cabin and walked through a stand of balsam fir and cedar to the shore.

Thornton is a chemist and an environmental engineer and the state of Minnesota's acid-rain-research coordinator. In one of its public-relations handouts, the National Coal Association has quoted Thornton as saying that no evidence exists that any Minnesota lake has become acid due to acid precipitation. When Thornton learned of the use of his statement, he was livid, for the quote is cast in a way that suggests that no acid-rain problem exists in Minnesota. Thornton vehemently disagrees with that implication, and he has data to support his disagreement. In fact, he brought me to Kane Lake to look at what almost surely is a dying body of water.

Early in 1982, Thornton had released a survey of Minnesota lakes that indicated that 512 to 967 lakes in eleven Minnesota counties already had lost most of their capacity to buffer further influxes of acid, and another 2,500 lakes, including 700 considered "major fishing lakes," were threatened by acid rain. Already, there were some indications that acid precipitation was leaching toxic metals into some of the state's drinking-water systems.

Small, undistinguished Kane Lake was one of the most sensitive of those sampled. Thornton would not suggest that Kane Lake *will* become acidic, just that the Scandinavian experience with acid lakes pointed strongly in that direction. Nor would he attempt to guess how long the lake might have: one year, five years, fifty years? No one knows.

But it is to the shores of Kane Lake, or another like it anywhere on the Canadian Shield or in any other sensitive area, that this story finally leads. Thousands upon thousands of lakes dotted across the planet that are sensitive to acid rain are still relatively healthy. So far, only a small fraction of sensitive waters have been damaged. But despite the industrial lobby's attempt to suggest otherwise, the threat is very real. As Thornton puts it, even though no acidified lakes have been found in Minnesota, "people don't understand that it's a *titration* process."

In other words, all over the globe, lakes are being subjected on a massive scale to the same sort of small-scale test-tube process that a chemist uses to measure alkalinity in a solution—adding acid drop by drop to be neutralized by the residual alkalinity. As long as any alkalinity remains, the pH will be stable. The lakes, of course, are much larger samples, but then rain and snow clouds are much larger sources of titration.

For Minnesota and the canoe country, Thornton's 1982 report concludes, "Continued loadings at these [current rain acidity] or higher levels will almost certainly cause detrimental chemical changes in at least the most sensitive ecosystems. At this time it is not possible to predict the length of time needed before more noticeable changes in lake chemistry occur."

Kane Lake may be one example of this large-scale titration. Data from 1951 indicated an alkalinity of 7.5 parts per million. The 1980s measurements show an alkalinity of only 3 parts per million. (To compensate for any possible error owing to changes in sampling technique, Thornton has calculated that the old alkalinity may have been as low as 5.3 parts per million—still well above the current level.) The decline in alkalinity in one body of water, Thornton insisted, is not necessarily cause for alarm—there could be a number of explanations, such as seasonal variations or sampling error. However, his data show that more than 80 percent of 209 low-alkaline lakes sampled in the region show significant declines in their capacity to buffer acid. "It's much more of a valid comparison because of the larger number of data points," he said. "At that point, we became concerned, so we did analyses of other chemical information, and this showed *other* chemical signs of the precursors to acidification." Those chemical signals, according to Thornton, include a shift toward sulfate as the dominant ion in the sensitive waters, whereas bicarbonate is usually the dominant ion in healthy waters.

He added, "Looking at all the chemical data, looking at the changes in alkalinity we've seen, looking at the pH of precipitation in northern Minnesota, looking at the similarity of this area to Scandinavia before lakes went acid, it is very suggestive that the titration process has already begun here."

The conclusion is unavoidable. Although there are no acidified lakes in Minnesota, although thousands of lakes in Scandinavia still have not been destroyed, although the Canadian lakes and streams that have been critically acidified represent only a tiny fraction of that nation's threatened water resources, and although in the Adirondacks and the rest of the northeastern United States the great majority of waters remain healthy, the titration is continuing. Whether the acidity of rain is currently increasing becomes almost a moot point, for the current levels are too high. For state, provincial, or national governments to allow industries to *increase* emissions is profoundly irresponsible. For industrial polluters to stall and obfuscate and even to pressure politicians for *weakening* current standards goes beyond poor corporate citizenship to villainy. For the very government officials charged with enforcing the environmental law of the land to throw their considerable weight behind the industrial lobby, and for those same officials to ignore the tidal wave of alarming data from the scientific community and instead concentrate on the trickle of information that tends to cast doubt is thoroughly despicable.

However, by early 1983, the Edison Electric Institute was still bent on convincing the nation's decision makers that acid rain really wasn't yet a proven threat. In January, the newsletter *Inside E.P.A.* reported that the Institute had hired "a high-powered New York–based public relations firm to engineer a national campaign against acid-deposition controls." The PR firm was to develop an organization called "The Alliance for Balanced Energy Solutions," with one mission—to stall acid-rain regulations. Membership fees would range from $10,000 to $50,000. Among the points the alliance would make were that "there is serious and honest doubt about the history, depth, nature, and trend of the effects of acid deposition"; that "there is sound reason for uncertainty about the extent of damage cause by acid deposition and the effectiveness of many proposed solutions''; and that ''the extraordinary economic impact of quick-fix solutions is not justified by the proven seriousness of the problem.''

Thornton and I drove from Kane Lake and along the north shore of Lake Superior toward the canoe country. The rocky shore is ribboned with small, fast-flowing, clear streams, almost all of which are trout streams. There is romance in the river names: the Baptism, the Temperance, the Devil Track, the Knife, the Cascade.

From the harbor town of Grand Marais just a few miles south of the Canadian border, we swung onto the Gunflint Trail, then thirty miles up the trail onto a small side road that took us into the canoe country, to the Clearwater Lake landing. Clearwater is a long, slim lake that stretches eastward from the landing and is one of the few lakes—all in the easternmost corner of the canoe country—that drain into Lake Superior.

We paddled for a couple of hours against a stiff breeze, past high, eroding bluffs of slate with great jumbles of fallen rock at their bases. ''Rove Logan slate,'' Thornton calls it. The slate tends to dissolve more readily than the older granite to the west of here, and as we paddle, Thornton, shouts back over his shoulder that just maybe this dissolution will actually make Clearwater Lake *less* acid-sensitive than others, since the rock should yield alkaline matter.

We had driven all day to reach the canoe country, and by the time we reached the eastern end of Clearwater, twilight was near. We made camp on a small cove without portaging out of this portal lake where motors are allowed, for the next day we intended to haul the canoe up a crude, steep, rugged hiking trail nearby to a large pond that is full of brook trout. We swam, cast small spinners from shore for bass, and prepared supper. The wind was dead, the water calm, and Thornton pulled out of his backpack a large plastic squirt bottle full of clear liquid and labeled "pH Buffer Solution," then a tiny saltshaker and a green container of artificial lime. Thornton, a native Texan, shook a bit of salt onto his hand, transferred it to his mouth, squeezed off a long shot of the contents of the chemical bottle, then the lime. He passed the containers to me. A loon, drifting out on our little cove, began to yodel.

That night as we slept in a tent by the shore of the cove, a thunderstorm rolled in over the huge, high bluffs of Rove Logan slate that surrounded us. It woke us both, and we listened as the thunder approached from the south, the lightning strobing against the green tent walls. There was a moment of heavy wind, patters of rain, then the lightning booming and crashing around us, a few minutes of deluge, and, just as quickly as it had come, the storm was fading away, and gone.

A few days after our return, Thornton telephoned to tell me that a rain monitor at Hovland, Minnesota, along the shore of Lake Superior, had recorded the storm's pH at 4.4

Index

Index

Index

Index

Index

Index